CELESTIAL WONDERS IN REFORMATION GERMANY

Religious Cultures in the Early Modern World

Series Editors: *Fernando Cervantes*
Peter Marshall
Philip Soergel

Titles in this Series

1 Possession, Puritanism and Print: Darrell, Harsnett, Shakespeare and the Elizabethan Exorcism Controversy
Marion Gibson

2 Visions of an Unseen World: Ghost Beliefs and Ghost Stories in Eighteenth-Century England
Sasha Handley

3 Diabolism in Colonial Peru, 1560–1750
Andrew Redden

4 Sacred History and National Identity: Comparisons between Early Modern Wales and Brittany
Jason Nice

5 Monstrous Births and Visual Culture in Sixteenth-Century Germany
Jennifer Spinks

6 The Religious Culture of Marian England
David Loades

7 Angels and Belief in England, 1480–1700
Laura Sangha

8 The Laudians and the Elizabethan Church: History, Conformity and Religious Identity in Post-Reformation England
Calvin Lane

9 Religious Space in Reformation England: Contesting the Past
Susan Guinn-Chipman

10 Anglo-German Relations and the Protestant Cause: Elizabethan Foreign Policy and Pan-Protestantism
David Scott Gehring

11 John Bale and Religious Conversion in Reformation England
Oliver Wort

12 Religious Diaspora in Early Modern Europe: Strategies of Exile
Timothy G. Fehler, Greta Grace Kroeker, Charles H. Parker and Jonathan Ray (eds)

Forthcoming Titles

Jews and the Renaissance of Synagogue Architecture, 1450–1730
Barry Stiefel

Priestly Resistance to the Early Reformation in Germany
Jourden Travis Moger

Images of Islam, 1453–1600: Turks in Germany and Central Europe
Charlotte Colding Smith

Exile and Religious Identity, 1500–1800
Jesse Spohnholz and Gary Waite (eds)

Female Piety and the Catholic Reformation in France
Jennifer Hillman

Indulgences after Luther: Pardons in Counter-Reformation France, 1520–1720
Elizabeth C. Tingle

CELESTIAL WONDERS IN REFORMATION GERMANY

BY

Ken Kurihara

Routledge
Taylor & Francis Group
LONDON AND NEW YORK

First published 2014 by Pickering & Chatto (Publishers) Limited

Published 2016 by Routledge
2 Park Square, Milton Park, Abingdon, Oxfordshire OX14 4RN
711 Third Avenue, New York, NY 10017, USA

First issued in paperback 2015

Routledge is an imprint of the Taylor & Francis Group, an informa business

BRITISH LIBRARY CATALOGUING IN PUBLICATION DATA

Kurihara, Ken, author.
Celestial wonders in Reformation Germany. – (Religious cultures in the early modern world)
1. Astronomy – Religious aspects – Christianity. 2. Reformation – Germany.
I. Title II. Series
200.9'43'09031-dc23

ISBN-13: 978-1-138-66306-0 (pbk)
ISBN-13: 978-1-8489-3444-3 (hbk)

Typeset by Pickering & Chatto (Publishers) Limited

CONTENTS

ACKNOWLEDGEMENTS

This study is based on my dissertation, which I completed in January 2010. My interest in *Wunderzeichen* first originated in a research seminar with W. David Myers, who later became my mentor and helped to formulate my research plan. Without his warm counsel, guidance and checking of my manuscripts, I would not have completed my dissertation. When my initial research plans reached a deadlock, Wolfgang Mueller suggested considering the political aspect of *Wunderzeichen* discourses. As I came across the 1549 tract on the Braunschweig *Wunderzeichen*, I found that his advice was sound and changed the focus of my study accordingly. I also appreciated his meticulous checking of my dissertation manuscripts. I share my interest in the early modern discourses on supernatural phenomena with Elaine F. Crane, and I was very grateful that she participated in my dissertation committee. I also thank Richard Gyug and J. Patrick Hornbeck II, who kindly agreed to my late request for them to become secondary readers of the committee; they all provided me with warm advice and encouragement. Susan Wabuda had a strong impact on my dissertation's perspective, and her constructive criticism of my initial research plan compelled me to reconsider the focus of my project.

Much of the initial research was made possible by research grants offered by Fordham University's Graduate School of Arts and Sciences, and I am grateful for their generosity.

I cannot express how thankful I am to Jill Bepler of Herzog August Bibliothek in Wolfenbüttel, who warmly welcomed me as a guest researcher. Gerlinde Strauss and other staff members of the library extended every courtesy to me during my research in the city. Christian Hogrefe helped me to obtain copyright permission from the library.

The British Library, Bayerische Staatsbibliothek, Columbia University Library, Burke Library of the Union Theological Seminary and Cornell University Library also gave me the opportunity to examine valuable sources. I would like to express my deep gratitude, especially towards Seth Kasten of Burke Library. A number of important primary sources were obtained through the Thrivent Reformation Research Program, and I thank the director, Mary Jane Haemig.

This study would not have become a book if Philip M. Soergel did not introduce me to Pickering & Chatto. He also responded to my inquiry regarding sixteenth-century wonder books and kindly explained his views on the topic. Luka Ilic gave me useful insights on Flacian theology on original sin and provided me with various secondary sources. I am grateful also to Jürgen Beyer, who generously sent me his manuscript on the Lutheran lay prophets and gave me inspiration on the relation between *Wunderzeichen* literature and pamphlets on angelic encounters. I am also very much obliged to Robert Kolb for his warm encouragement.

I owe a huge debt to Philip Good and Janka Romero of Pickering & Chatto. My thanks also go to everyone at the publishing house who worked on the production of this volume.

Unfortunately, I cannot list the names of all of the friends who supported me during the long years of research, but my gratitude goes out to you. My family, especially my parents, Hiroshi and Yoshiko Kurihara, continuously encouraged me from Japan with both moral and financial support. My parents-in-law, Jun and Fujiko Tonegawa, relaxed my mind by inviting me to their home and providing additional kindness. Finally, I want to mention my wife, Noriko. She endured my long absences during my research trips, which started only three and a half months after our wedding; without her loving support, I could never have finished this study.

LIST OF FIGURES

INTRODUCTION: CELESTIAL WONDERS, CONFESSIONAL CONFLICTS AND APOCALYPTICISM

The Topic: *Wunderzeichen*

In sixteenth-century Germany, *Wunderzeichen* ('wonder-signs') could signify all kinds of wondrous phenomena observed in nature. According to Michaela Schwegler, the category of *Wunderzeichen* includes wonders that appeared in the sky, wonders related to grains (the raining of corn, or the miraculous finding of grains), wonders related to blood (the raining of blood, or blood which appeared in springs), and other kinds of mysterious incidents such as encounters with angels, prophetic children, ghostly apparitions and resurrections of the dead.[1] Hiob Fincel's *Wunderzeichen* (1556), the pioneer of early modern books on wonders, includes not only these phenomena but also monstrous births, urban fires, natural disasters and terrible murders. However, in many cases the term *Wunderzeichen* was used to signify unusual celestial phenomena: comets, irregular movements of the sun and the moon (such as eclipses and halo phenomena), novae, *Fewerzeichen* (the Northern Lights), and apparitions of various strange objects such as rods (*Rute*), crosses, funeral processions, Christ, angels, armies of soldiers, and animals. In this study, my focus is limited to these wonders that appeared in the sky. People in Lutheran Germany in the sixteenth and the early seventeenth centuries showed great interest in these phenomena, and the sheer number of prints concerning strange celestial signs published during this period suggests their enormous popularity. Gustav Hellmann lists nearly four hundred titles (including reprints) of celestial *Wunderzeichen* broadsheets and tracts which appeared between 1550 and 1599.[2] Yet these pieces are only a portion of the literature on this topic. If we add all the contemporary books and sermons which discuss celestial phenomena as divine wonders, this number must be doubled.

Broadsheets and Clerical Literature

Our general images of *Wunderzeichen* are largely based on the woodcut illustrations printed on broadsheets because many of them are available in modern published collections of German *Einblattdruck* (broadsheets) and a number are also printed in various books on early modern European culture. Thanks to the collections compiled by Bruno Weber,[3] Walter L. Strauss,[4] Dorothy Alexander (in collaboration with W. L. Strauss),[5] Wolfgang Harms (and his collaborators),[6] Christina Hofmann-Randall[7] and Axel Janeck (the exhibition catalogue of the Germanisches Museum in Nuremberg),[8] more than one hundred examples of *Wunderzeichen* broadsheets can be seen even without visiting the libraries in Zurich and Gotha, both known for their large collections of *Einblattdruck*. The illustrations in these prints, vividly depicting the dramatic appearance of strange phenomena and the bewilderment of the people seeing them, are eye-catching and stimulating. It is understandable that a number of studies in *Wunderzeichen* have used these illustrated broadsheets as primary sources.[9]

Yet constructing one's analysis solely based on these broadsheets contains several dangers. Since a broadsheet has only one page and much space is occupied by the illustration, the length of the attached texts is limited, and in many cases the texts do not contain much information about the circumstances of publication and the authors' intentions. Furthermore, the contents and the narrative style of these texts show remarkable similarity to each other because many are simply following an established format. Of course, we can grasp the general trend of *Wunderzeichen* literature based on these texts: fear of God's wrath and the imminent coming of the Last Judgement echo through almost all the *Wunderzeichen* broadsheets. However, to 'decipher' the specific meaning of each broadsheet, caution is needed; unless one examines the context of each print carefully, his or her analysis can easily fall into wild guessing.

This does not mean that we should avoid using broadsheets as sources. Alexandra Walsham rightly points out that the strange visions which people saw in the sky merit more attention because they 'provide us with a unique glimpse of their deepest fear and anxieties'.[10] Illustrations on broadsheets which depict these apparitions can serve as important 'windows' into the minds of the people in early modern Europe. Nevertheless, one should note that it is essential to examine the context of each *Wunderzeichen* report carefully and to consult with other primary source genres.

Indeed, broadsheets are only a part of *Wunderzeichen* literature. Sixteenth-century people also discussed these strange phenomena in *Flugschrift* (news pamphlets), *Wunderbuch* (wonder books), astrological tracts, sermons, devotional books, scientific works and personal letters. Among these, one type of source received the particular attention of scholars: *Wunderbuch*, the catalogues of *Wunderzeichen* news compiled by Hiob Fincel, Conrad Lyscothenes, Caspar Goltwurm and others. These works, which first appeared in the 1550s, immediately won great popularity among Lutheran readers and shaped the direction for the later development of *Wunderzeichen* discourses. Several articles have been written to analyse the contents and the sources of these *Prodigiensammlungen* (collections of prodigies) and the lives of their authors.[11]

Among the three compilers of wonder books named above, only Caspar Goltwurm was a pastor. Nonetheless, when we check the publication data of other genres of *Wunderzeichen* literature, the first thing we notice is that the majority of the authors of these materials are the clergy. Lutheran preachers showed great interest in strange celestial phenomena and often incorporated news of *Wunderzeichen* in their sermons and other writings. Based on Christ's prophecy of the world's end in Luke 21 and Matthew 24, these celestial phenomena were regarded as warning signs of the imminent coming of the Last Judgement, and they offered support for the clergy's eschatological world view. In addition, since it was generally believed that *Wunderzeichen* signified God's wrath towards unrepentant sinners, news of strange phenomena was used as a convenient tool to terrify people and to press them to turn to God. As we shall see in Chapter 2, most of the prominent Lutheran theologians in the latter half of the sixteenth century mentioned *Wunderzeichen* in their writings, and a number of them showed their extensive knowledge about contemporary news of strange phenomena.

However, compared with the wonder books of Fincel and others, the clergy's discourses on *Wunderzeichen* have not received much scholarly attention, and until very recent years there had been only a few systematic studies that focus on the clergy's involvement in the production of *Wunderzeichen* literature.[12] Among these studies, Philip M. Soergel's *Miracles and the Protestant Imagination: The Evangelical Wonder Book in Reformation Germany* (2012) is by far the most outstanding contribution to this field in recent years. Focusing on the history of the Lutheran interest in prodigies, it analyses the theological world view underlying sixteenth-century wonder books, including those compiled by the clergy (we will return to his monograph later).[13] Another inspiring example is C. Scott Dixon's article on sixteenth-century Lutheran astrological literature. Dixon points out that a considerable number of respected ministers published or commented on astrological predictions. He then analyses the influences of Lutheran eschatology and the clergy's frustrations with the slow progress of the Reformation in these astrological writings. This analysis reveals that the 'root of the Lutheran almanacs' in the late sixteenth century was the clergy's 'sense of frustration and failure, the conviction that the preaching of the gospel was not enough to indoctrinate the teachings of the church or the fear of God'.[14] These pastors turned to 'preaching of the stars' as a different medium of sermons and tried to admonish people by almanacs. Dixon's keen observations can directly be applied to the Lutheran clergy's *Wunderzeichen* literature in general, but such a study has not been written yet.

This relative neglect of the clergy's involvement in *Wunderzeichen* is to some degree understandable. Though a number of theologians published tracts, books and sermons focusing on *Wunderzeichen*, in many cases the clergy's references to *Wunderzeichen* are scattered in various religious publications such as postils, regular sermons, *Busspredigten* (sermons calling for repentance), sermons on the second Sunday of Advent, and devotional books on the Last Judgement and other topics. To comb through these works and to pick up references to celestial wonders one by one is time-consuming. In addition, in order to comprehend why the clergy were interested in these strange phenomena and what they intended to teach through *Wunderzeichen*, one must first understand the complex world of Lutheran eschatology because the apocalyptic world view was the backbone of *Wunderzeichen* discourses in the sixteenth century. Moreover, one should also examine the theological positions of the authors and see how they viewed the world surrounding the Church in their days. Considering the stormy condition of the Lutheran Church in the latter half of the sixteenth century, filled with complicated theological controversies, to grasp an author's theological position is not an easy task. Thus it is understandable that the relations between the clergy and *Wunderzeichen* literature have not yet been fully explored.

However, considering the fact that the clergy were the intellectual leaders of sixteenth-century communities and their strong influence on people's understanding of prodigies, we need to pay more attention to this topic. Moreover, a systematic study of the clergy's *Wunderzeichen* writings will also make a significant contribution to a recent trend in the history of science, which focuses on the vivid interactions between theology and science in early modern Europe, as early modern astronomy had strong connections with *Wunderzeichen*. In her work *Kepler's Tübingen* (1998), Charlotte Methuen states:

> It is only in understanding how science, theology, and other expressions of culture interact with one another to confirm, but also to change, that culture and each other, that it is possible to understand how modern western intellectual and scientific standards came into being.[15]

In her book, she examines the theological views on nature among the theologians and the scientists of Tübingen University in the latter half of the sixteenth century and their impact on the young Johannes Kepler. Methuen emphasizes the significance of the educational influences of Philipp Melanchthon, who regarded astronomy as the most important of the mathematical science because he believed that the perfect order of the movements of the heavenly bodies signified God's will and plan.[16] She carefully examines the theological views on nature of Jacob Heerbrand, Kepler's teacher of theology at Tübingen, which showed the marked influence of Melanchthon. *Wunderzeichen* are not forgotten here. Shortly after the great comet appeared in 1577, Heerbrand delivered a sermon warning people that this comet was a sign of God's wrath and punishment, and Methuen devotes four pages to summarizing the contents of this sermon.[17] As she points out, despite his sceptical attitude towards astrology, Kepler's astronomy teacher Michael Maestlin (who was also a cleric, because he had once served as a pastor) did not hesitate to declare the comet as God's call for repentance.[18] All of these theological/scientific discourses crystallized Kepler's self-identification as a 'priest of the book of nature' who 'interprets the message of the heavens to the glory of God'.[19] For Kepler, the devotional aspect of astronomy was not mere rhetoric but the very core of his thought. Methuen's book not only provides background information of *Wunderzeichen* literature but also shows the possibility of how studying the interactions between theology, celestial wonders and astronomy can lead to a new dimension in the history of science. For this purpose, more study about theologians' writings on celestial phenomena is desired.

Lutheran Apocalypticism

Since Lutheran eschatology is the key to understanding the relation between the clergy and *Wunderzeichen* discourses, here we should examine the best study of Lutheran apocalypticism available in English: Robin Bruce Barnes's *Prophecy and Gnosis: Apocalypticism in the Wake of the Lutheran Reformation* (1988), which offers a bird's-eye view of the development and decline of apocalyptic thinking of German Lutherans. According to Barnes, apocalypticism was not only a part but the very core of Lutheranism until the Thirty Years' War. From Martin Luther in the 1520s to Jakob Böhme in the 1620s, Lutherans were almost 'obsessed' with the idea that the end of the world was imminent.[20] Apocalypticism had existed since the time of the early Church, but Luther added a new dimension to the traditional eschatology. Since the Gospel was restored and Antichrist was revealed in the papacy, for Luther and his followers, it was clear that the Day of the Last Judgement could not be far away; the final cosmic struggle between the Church of Christ and the Church of the devil/Antichrist had started. As the years passed, Luther became deeply disappointed by the slow progress of the Reformation and did not hide his fear that the German people, blessed with the preaching of the Word of God for many years, were now missing their last chance to turn to Christ. This gloomy world view was inherited and developed by subsequent generations of reformers. The conditions surrounding their Church were hardly encouraging during the years following Luther's death. The Schmalkaldic War (1546–7), the Interim Crisis (1548–52), the theological controversies which split the Church, the advance of the Counter-Reformation, the spread of Calvinism and the shadow of religious wars – these threats strengthened the siege mentality of Lutheran theologians and intensified their apocalyptic concerns. Quoting hundreds of primary sources, Barnes shows how strongly Lutheran theologians during the sixteenth and the early seventeenth centuries were obsessed with the deep fear that the Last Judgement would come while most people in the world were still misled by the devil.

This is the background of Lutheran clerical writings on *Wunderzeichen*. Based on various sixteenth-century theological and astrological discourses, Barnes spends many pages describing Lutheran enthusiasm for *Wunderzeichen* and astrology and its connection with the church leaders' eschatology. 'Indeed', he states, 'by 1560 attention to the unusual in nature had become nothing less than an obsession'.[21] Barnes's arguments show that a number of prominent theologians, including Philip Melanchthon, Andreas Musculus and Simon Pauli, showed profound interest in *Wunderzeichen*, and the eschatological interpretation of celestial phenomena had a strong influence on the scientific discourses of Tycho Brahe and other astronomers.[22] It is Barnes's achievement to highlight the dynamic interactions between Lutheran theology, apocalyptic world views and the conditions of sixteenth-century society, a topic that had largely been neglected by historians. His study shows that people's interest in *Wunderzeichen* was by no means 'medieval superstition' or the result of ignorance; it belonged to the mainstream of the orthodox theology of Lutheranism, and many leaders of the Church actively promoted it. Barnes also successfully shows the immensely rich diversity of Lutheran eschatological discourses. Though they shared the same conviction that the world was coming to an end, Lutherans expressed their belief in a variety of ways. Some tried to decode Scripture passages to predict the timeline of the End of Days, while others turned to Kabbalah and other 'ancient' knowledge to comprehend the mysteries in Revelation. Many eagerly observed the movements of the heavenly bodies and looked for apocalyptic messages in the 'preaching of the stars'.

However, and despite these thought-provoking contents, there is one topic which is not covered to a great extent in Barnes's book. Between 1547 and 1580 the Lutheran Church was largely divided between 'Philippists' and 'Gnesio-Lutherans', and neither group hid their suspicions of the other. Was there any relation between these theological controversies and apocalyptic *Wunderzeichen* discourses? Barnes briefly explains the theological differences between the two groups, but he later assesses that 'Gnesio-Lutherans, Philippists, and other factions did not differ in their central eschatological teachings' and 'here the theological issues can be safely ignored' because '[w]hat is important is that this sort of disagreement helped to fuel the apocalyptic sensibility that was already strong within Lutheranism ... The resulting confusion only heightened the common expectation of imminent Judgment'.[23] Barnes rightly recognizes that these theologians in conflict shared the conviction that Judgement Day was at hand and that they saw the controversies plaguing their Church in an eschatological light. However, by generalizing these authors' voices, one can overlook the rich diversity of the apocalyptic conviction expressed in the *Wunderzeichen* discourses and the interesting dramas hidden behind individual works. As will be seen in subsequent chapters, several Lutheran theologians used *Wunderzeichen* as polemical weapons against theological opponents. The clergy, like Christoph Irenaeus, employed stories of celestial wonder to convince readers that strange phenomena appeared as warnings against the false teachers of the End Times. In the case of Irenaeus, the false teachers were Calvinists, Synergists and later the formulators of the Formula of Concord. In such cases the theological issues were at the very centre of these discourses and inseparable from the authors' world view. Borrowing Ronald K. Rittger's words, it is possible to state that '[a]s was true of nearly every aspect of culture and human experience in early modern Germany', celestial phenomena 'became confessionalized'.[24] Considering these points, interactions between theological controversies, apocalypticism and *Wunderzeichen* discourses deserve more scholarly attention.

Of course, stories of wonder are not the main focus of Barnes's study, and the omission of a few polemical works does not diminish the immense value of his work. Barnes's aim is to give a general overview of the world of Lutheran eschatology and explain the characteristics of End Time discourses, and with the use of 500 primary sources, he successfully accomplishes his goal. However, if we want to comprehend the roles *Wunderzeichen* fulfilled in sixteenth-century society and their relations with Lutheran religious culture, we must examine the complex interconnections between confessional conflicts and celestial wonders.

Prodigies, Confessional Conflicts and the Perception of Sins

The relationship between religious turmoil and the world of wonder is discussed in several recent works on early modern prodigies. Early modern monstrous births have received more scholarly attention than celestial wonders, and some of these studies give useful insight to *Wunderzeichen* research. Jennifer Spinks's *Monstrous Births and Visual Culture in Sixteenth-Century Germany* (2009) focuses on the history of visual representations of misbirths throughout the sixteenth century (including the pre-Reformation period) and explores the diversity of people's perceptions of deformity and the meaning attached to individual incidents. Spinks points out that before the Protestant Reformation, monstrous births were not necessarily depicted in a negative light and were sometimes interpreted as positive omens. However, Luther's and Melanchthon's use of the stories of the Monk-Calf in Saxony and the infamous Papal Ass brought 'a decisive shift in attitude, in which interpretation and representation became not only more polemical – and particularly anti-papal – but took on a notably apocalyptic aspect'.[25] This fearful image of misbirths as an apocalyptic symbol was popularized through the illustrated wonder books by Fincel and Lycosthenes and the broadsheets juxtaposing monstrous births with other wonders and disasters. Spinks points out that Protestants were not the only group who used this topic for confessional propaganda. In his satiric print, Johann Nas, the Franciscan polemist of Ingolstadt, incorporated various images of deformed children born in Germany to mock the false nature of Luther's Church. The last primary source analysed in Spinks's study is the tract relating news of a three-headed baby in Brandenburg in the 1590s who verbally warned of imminent divine punishment. Though the text was not produced for specific propaganda purposes, we cannot grasp the full meanings of the news unless we understand '[t]he interaction of Lutheran, Calvinist and Catholic confessions in a period of intense uncertainty, marked by new anxieties about the Turkish threat', as the background of the report.[26] Thus Spinks's book reveals that negative representations of misbirths were established and developed in reaction to the confessional conflicts brought about by the Reformation.

Julie Crawford's *Marvelous Protestantism: Monstrous Births in Post-Reformation England* (2005) also analyses the various social and religious functions fulfilled by the printed reports of misbirths in early modern England. Crawford's careful examinations of the social backgrounds of each report of monstrous births reveal how the stories of deformed babies were used to support the Protestant activists' campaigns against luxurious fashions, sexual immorality and other vices, mainly those committed by unruly women. Some messages, such as the Puritan minister William Leigh's sermon on the birth of the double-faced child in 1613, could contain political implications. In his sermon, quoting the news of the misbirth, Leigh condemns the Protestants' 'equivocation' and accuses that while they believed the Scripture's teaching, people 'do not live the life of the Gospel'. This sounds like a common moral admonition, but as Crawford points out, it could be a criticism against James I, whose equivocal dealings with Catholics won notoriety among the Puritans.[27] However, the cases more directly related with ongoing confessional conflicts are the two births of headless babies which occurred during the 1640s. One is the case of a woman who delivered a headless baby after declaring that she wanted her child to 'be born without a head, then to be signed with the signe of the Cross'. The other is about a recusant woman who stated that 'I would rather bear a child without a head then a roundhead' and ended up with the same result.[28] Understandably, the former report was used as propaganda against sectarianism while the latter was circulated to buttress Parliament's cause. Thus the bodies of deformed babies were interpreted from the perspectives of confessional politics and actively exploited to promote political agendas.

Crawford's focus is on the patriarchal attempts to control women's behaviours observed in these reports of wonder, and she states in one chapter: '[i]n fact, I suspect that some original determinations of deformed and stillborn births as "monsters" were based on perceptions of their mother's sins rather than on the actual forms of the births themselves'.[29] Her sharp insight is also inspiring for us when we consider the meanings of *Wunderzeichen*. In the case of monstrous births, perceptions of their mothers' sins made people see monstrosity in the deformed shapes of the poor babies; in the case of celestial wonders, in a sense, the clergy's perceptions of human sins prevailing in society made celestial phenomena wondrous signs of God's wrath. For Irenaeus, it was the sin of false prophets, the erroneous teachings of his theological opponents, which brought various strange phenomena in the sky. Thus discourses on prodigies are inseparable from the perception of sin in early modern Europe.

Here let us return to Soergel's *Miracles and the Protestant Imagination*, which focuses on the theological background underlying the Lutheran wonder books that flourished between the 1550s and the 1600s. As his propagandist use of deformed animals indicates, Luther was fascinated with natural signs. However, he did not indulge in speculation about the pessimistic meanings of wondrous signs apart from God's Word; rather, he stressed the benevolence of God working in nature which preserves this world, despite the sinfulness of humankind. The followers of Luther were less balanced than their teacher. The wonder books, which emerged from the deep uncertainty during the years after the Schmalkaldic War, were essentially preaching the Law, which emphasized human depravity and God's wrath and punishment on human beings' transgressions of divine commandments:

> Each event, these authors [of wonder books] warned, was explicit punishment for a particular vice. In this way, the evangelical wonder book attempted to drive home a vital distinction long central to Reformation teaching: the relative differences between Law and Gospel.[30]

Soergel believes that what was underlying such a tendency was 'a deep uneasiness at the heart of a maturing Reformation, an uneasiness prompted by the growing realization that the teachings of faith alone were insufficient to ensure social discipline'.[31]

Indeed, stories of wonder as warning of divine wrath were self-disciplinary tools which offer 'a remedy for the threat of antinomianism'. Wonder books 'reinforced the vital role, identified by Melanchthon and other early evangelicals, of the Law as a set of proscriptions that not only convicted the unrighteous but also bound the justified'. Unlike the early modern popular stories of Catholic miracles which commemorated God's mercy, Protestant wonder tales were 'almost universally negative in tone' and cannot hide 'the unrelieved pessimism that ran through the new genre'.[32] This gloomy tendency culminated with the wonder books of Irenaeus, whose theology on original sin centred on the total depravity and corruption of human nature which lost the image of God through the Fall. Soergel believes that this trend gradually changed around 1600, as the territorial Church assumed the corrective functions that Lutheran pastors had tried to exercise through the relentless preaching of the Law. The clergy became more eager to provide spiritual consolation for those who were in challenging situations (such as natural disasters), and people became more open to empirical observation of natural phenomena, no longer bound by the traditional reading of wonders. Thus Soergel clarifies that what constitutes the theological foundation of *Wunderzeichen* discourses was preaching of the Law, God's condemnation of human beings' wrongdoings, which was supposed to lead people to turn to God's mercy. This overemphasis of the Law imposes serious limitations to the message of wonder, as we shall see later.

The Structure of this Study

There are several further interesting studies that discuss the relation between prodigies and religious turmoil, such as Alexandra Walsham's *Providence in Early Modern England* (1999), but the above texts have shown the direction our study should take.[33] The points which should be focused on are confessional conflicts, apocalypticism and the theological perception of sin of humankind as the formative forces of *Wunderzeichen* discourses. If *Wunderzeichen* discourses were products of confessional struggles, what kind of functions did these texts fulfil in the process of the confessionalization of sixteenth-century Germany, including theological controversies? This is the topic to be pursued in this study. Since the interaction between confessional struggles and *Wunderzeichen* literature is the main point to explore, the primary sources of my study are the Lutheran clergy's publications on celestial wonders, such as sermons, postils, tracts and wonder books. One point to note is that when we analyse *Wunderzeichen* works by the clergy, we have to examine the context of each text carefully. Especially important is comprehend which theological positions the authors were taking and to consider for what purpose they were using *Wunderzeichen* in their arguments – were they quoting the stories of wonder simply to edify people and urge them to repent? Or was polemical intention hidden behinds their words? At a glance, most of the *Wunderzeichen* works seem to repeat the same message endlessly: 'the Last Judgement is at hand' and 'repent or disaster!' However, if we carefully dissect the contexts of individual works, we can find that these callings are not mere stock phrases; they also hide the various concerns and agendas of the Lutheran clergy.

Chapter 1 explores the world of *Wunderzeichen* in general, in order to grasp what kinds of media people used to obtain information about celestial wonders, what celestial phenomena were regarded as *Wunderzeichen*, and how contemporary people interpreted their meanings. This chapter resembles a more traditional *Wunderzeichen* study. Besides examining several writings by Lutheran ministers, a number of broadsheets and tracts are also used as primary sources. In the latter half of this chapter, I shall argue that the immense popularity of *Wunderzeichen* literature is based on the fact that these discourses directly reflected the people's concerns about changing climatic conditions, the danger of foreign invasions and the rapid development of confessional conflicts.

Chapter 2 presents a general overview of the Lutheran clergy's involvement in the production of *Wunderzeichen* discourses. Readers will be surprised by the list of theologians who showed interest in strange celestial phenomena, as the names include almost all the leading theologians of the day. After examining the academic, spiritual and emotional factors that stimulated the clergy's interest in *Wunderzeichen*, we will analyse the religious functions fulfilled by the clergy's *Wunderzeichen* publications in the sixteenth-century community. One important point to note is the role of celestial wonders as *Gesetzpredigt* (sermon of the Law) – this is the key to comprehending the nature of the clergy's *Wunderzeichen* literature. Also noteworthy is the clergy's understanding of celestial wonders as *Beiprediger* (co-preachers) sent by God. Preachers claimed that because people had refused to listen to their human preachers' warnings, God sent the mysterious signs as special messengers to convey his own warning. Behind this idea, there was most likely the emotional need of the clergy to reassert their identity and authority as servants of God's Word in the midst of the anticlerical atmosphere.

Chapters 3 through 6 focus more on the context of specific *Wunderzeichen* works. The work discussed in Chapter 3 is a report of strange celestial visions published by Matthias Flacius Illyricus during the Interim Crisis. Flacius published this tract (written by another author, Nicholas Medler) to employ the case as political propaganda against the religious policy of Emperor Charles V and the Wittenberg theologians who yielded to Charles's political pressure. With the help of other contemporary *Wunderzeichen* works, the political meanings of the images that appeared in the sky will be clarified.

Chapters 4 and 5 examine the wonder books written by Christoph Irenaeus, who once served as the court preacher of Weimar but later went into exile due to his theological understanding of original sin. He was an enthusiastic collector of stories of wonders, and he published four wonder books that focus on celestial *Zornzeichen* (signs of wrath), comets, monstrous births and natural disasters, especially floods and storms. The very purpose of publication of these texts was to attack the 'false preachers' in his days (which meant his theological opponents), and Irenaeus used *Wunderzeichen* stories to catch people's attention and then lead them into his theological discussions. Chapter 4 looks at his books in the 1560s to show how he developed his rhetorical skill to employ stories of prodigies for his specific confessional agendas, while Chapter 5 focuses on Irenaeus's polemical works in the 1570s and 1580s, composed after he was expelled from Saxony due to his theology on original sin. Throughout these years in exile, he devoted his life to attacking the doctrine of original sin in the Formula of Concord and denouncing its formulator, Jacob Andreae, as the arch false teacher of his days. This latter chapter reveals how Irenaeus used stories of comets and monstrous births to condemn Andreae's work for the Concord and to promote his theological stance on original sin.

Chapter 6 is about how Jacob Andreae, the chief architect of the Formula of Concord, used *Wunderzeichen* for his pastoral care. Though Andreae mentioned celestial signs in some of his works, unlike Irenaeus he seemed hardly enthusiastic about this topic. It seems he feared that careless references to celestial phenomena would encourage people's interest in astrology, an art he denounced as idolatry. However, when he judged that it would be useful to guide the laity, Andreae did not hesitate to incorporate astrological imagery and news of celestial signs in his messages. This suggests that Andreae's use of *Wunderzeichen* news was motivated by his sincere pastoral concerns and reflects his effort to indoctrinate common people in the most comprehensive way for hearers/readers.

Chapter 7 analyses *Wunderzeichen* discourses during the two decades preceding the outbreak of the Thirty Years' War. The doctrinal unity of the Lutheran Church was officially achieved by the Concord, but the following period was characterized by growing tensions between Catholic powers and Protestant states. Moreover, during this period the Lutheran Church had to confront challenges coming from within: the rise of proto-pietism and the spread of various occult beliefs. This chapter examines whether these difficult factors of the age affected the contents of *Wunderzeichen* works. The endlessly repeated warning of imminent divine wrath finally came true with the comet in 1618 and the outbreak of the Thirty Years' War, but ironically this also became the beginning of the decline of *Wunderzeichen* literature. The possible causes of the decline of these discourses after 1618 are discussed.

While the period covered in this study is basically between 1546 (the year of Luther's death) and 1595 (the year of Irenaeus's death), the period between 1595 and 1618 (the outbreak of the Thirty Years' War) is also discussed in Chapter 7. The years between 1546 and 1618 can be characterized as the period of the development of confessionalization, the age that established and solidified confessional identities on the theological, political and cultural levels. The heyday of *Wunderzeichen* literature and Lutheran apocalypticism largely coincides with these seventy-two years.

1 EXPLORING THE WORLD OF *WUNDERZEICHEN*

The *Mansfeld Chronicle* (1572), compiled by the Mansfeld deacon Cyriakus Spangenberg, vividly depicts the difficult environment people had to confront while living in the 'Little Ice Age'. Deterioration of climatic conditions, severe storms, repeated floods, inflation, pestilence and fatal accidents regularly appear in the entries. As though putting lively accents into this gloomy narrative, Spangenberg inserts stories of wondrous incidents that occurred in the region. These include reports of misbirths, divine punishments on disbelievers and a mysterious spectre seen in the mountains, but the predominant types of wonder are celestial wonders: 'People saw the three suns'; 'some people ... saw a cross in the sky, followed by a hand holding two Turkish sables'; 'some say that people saw two blood-red men in the sky'; 'in many places in the land of Sachsen, the nightly sky was seen ... as if it is really burning'; 'a blood-red beam was seen ... After that a cloud stood in the sky, [which looks] like a bier for the dead with a white cloth and black cross'. Such entries frequently appear in the pages.[1] Why did the prominent clergy of the city show such an interest in these phenomena? How did Spangenberg and his readers interpret these incidents? Some of the wondrous phenomena recorded in the chronicle took place in distant lands such as Lübeck – how did Spangenberg obtain information about such incidents?

In order to find answers to these questions, this chapter explores the world of *Wunderzeichen* literature between 1546 and 1595 so that we can grasp what kinds of celestial phenomena were called *Wunderzeichen*; how the media heard about these strange phenomena; how they interpreted the meaning of *Wunderzeichen*; and why the news of these wondrous incidents drew their attention. First we will examine six types of popular media for *Wunderzeichen* news in order to comprehend how stories of wondrous phenomena were circulated among people. Then *Wunderzeichen* will be classified into four categories, and the various examples of celestial phenomena and the interpretations attached to them by contemporary authors will be explored. A preliminary examination shows that virtually all *Wunderzeichen* literature carries the same message: 'Repent, or disasters will come'. The climatic, social and political conditions of German society in the second half of the sixteenth century show that people had ample reason to fear that various catastrophes would soon befall Germany. Indeed, *Wunderzeichen* reflected people's widespread fears of possible catastrophe and their desire to know what they could expect for the future. This inner concern is the background for the clergy's use of celestial phenomena in their writings.

Media for *Wunderzeichen* News

The media for *Wunderzeichen* news can be classified into the following six categories: *Einblattdruck/Flugblatt* (broadsheets); *Flugschrift* (pamphlets); *Wunderbuch* (wonder books); *Prognosticon* or *Practica* (astrological predictions); sermons and other devotional works written by the clergy; and oral communication.

The most popular and readily available media was *Einblattdruck*, a one-page broadsheet usually consisting of a title, a woodcut illustration, a brief description of the *Wunderzeichen*, the author's comments and a prayer for God's mercy. The author often attached his own interpretation of the phenomenon and admonished readers to keep the lessons in mind. The price of a broadsheet is difficult to ascertain, but Bruno Weber estimates that one broadsheet cost at most 1–2 Kreuzer or 4–8 Pfennig in the sixteenth century.[2] Usually a large woodcut illustration depicting the phenomenon is printed at the top of the sheet. In many cases, the depictions are fairly in accord with descriptions in the text. Some texts do not explain the phenomenon in detail but simply state that 'a strange phenomenon, as depicted above, appeared in the sky', and such examples show a strong interrelation between the text and the visual image in *Einblattdruck*. Since the average literacy rate of sixteenth-century Germany is estimated at around 5 per cent (though it could reach 30 per cent in urban areas), the role of visual images was significant in mass communication.

While the authors of these *Einblattdruck* usually remain anonymous, some texts carry authors' names, and some of these were of prestigious status. For example, a Wittenberg broadsheet about the appearance of three suns and mysterious rainbows in March 1551 is signed by no other than Philip Melanchthon.[3] Georg Henisch (1549–1618), the instructor of St Anna Gymnasium in Augsburg who is known as the editor of the first German dictionary (1616), discusses the astrological influences of the 1580 comet in an Augsburg *Einblattdruck*.[4] Some historians assume that the report of strange eclipses and apparitions of a large army in the sky over Württemberg (1578) was written by Johann Fischart, one of the most famous authors of the German Renaissance.[5] The printer of a broadsheet reporting a strange celestial fire in France (1554) claimed that the original account was written by the French court astrologer Michel de Notre Dame, better known as Nostradamus.[6] At least one *Wunderzeichen* broadsheet was illustrated by Jost Amman, a prominent Nuremberg artist known for his *Book of Trades* illustrations, although most broadsheet woodcuts were made by lesser skilled artists.[7] Notable authors and elaborate illustrations attached to their texts suggest that the traditional image of *Wunderzeichen* broadsheets as cheap, unintelligent products of popular culture needs correction.

A similar media to *Einblattdruck* is *Flugschrift*, an unbound news pamphlet or tract with multiple pages. The structure of *Wunderzeichen Flugschrift* is basically the same as that of the broadsheet: a sensational title on the front page, a description of the phenomenon, the author's comments, and a prayer to God as a conclusion. However, since there is no one-page limit, an author could amplify the content of his text by reporting on more than one *Wunderzeichen* case or other kinds of eye-catching incidents in a single tract. A number of these pamphlets report on more than two or three cases and quote historical examples of *Wunderzeichen*, Scripture verses, words of ancient sages or lengthy moral teachings to enhance the commercial value of the text. While an illustration often appears on the title page, in many cases it is only for decoration rather than an accurate depiction of the phenomenon in the text. Though the price of *Flugschrift* depended on the size of the booklet, it is estimated to be around 20 Pfennig.[8] Many of these works remain anonymous, but authors' names do appear more frequently in *Flugschrift* than in *Einblattdruck*. Some texts also carry the date of publication or the date of composition, and this enables us to deduce the circumstances of the publication. One example is *True Report and Short Account of Great and Dreadful Wunderzeichen ... Appeared in the Sun* (*Warhafftige Bericht Vnd Kurtze Beschreibung des Grossen vnnd Erschrecklichen Wunderzeichens ... an der Sonnen begeben*; published in Hof, 1584), a six-page tract about a strange incident in the village of Zeyern on Easter Sunday, 19 April 1584. According to the author, Wilhelm Pletzlein, the pastor of Zeyern, fire suddenly sprang from the sun and fell onto the earth, followed by bread falling from heaven.[9] The date of the publication, printed at the end of the tract, is 30 April. This means that this village pastor composed the account, sent it to a printer in Hof, and published it within eleven days of the incident – an impressive speed, considering that Hof was not a publication centre in those days. Such an example shows that *Flugschrift*, which did not need elaborate illustrations, was an express media that enabled quick circulation of *Wunderzeichen* news.

When stories printed in broadsheets and tracts were assembled into one volume, they became a *Wunderbuch*, a popular literary genre that newly emerged in the mid-sixteenth century. Wonder books usually catalogue stories of strange phenomena in chronological order or classify them by types of phenomena, for example eclipses, comets, rainbows and strange images. The first *Wunderbuch* in Germany was compiled in 1532 by Joachim Camerarius, a close friend of Melanchthon, but this work was written in Latin and not for a popular audience.[10]

The pioneer work of popular *Wunderbuch* is *Wunderzeichen* (1556) compiled by Hiob Fincel, a Lutheran physician and professor of philosophy at Jena University.[11] The book lists all kinds of strange events after 1517, including not only celestial wonders but also monstrous births, strange behaviours of animals, manifestations of devils, and various disasters such as fires, storms, earthquakes and floods. Many of these stories were directly taken from *Einblattdruck* and *Flugschrift*. Fincel also explains what kinds of misfortunes followed after these portents appeared and asserts that the recent increase of these strange signs clearly indicates that the Day of the Last Judgement is quickly approaching. The book won such great popularity that Fincel published a second (1559) and a third volume (1562) of *Wunderzeichen* to update his story collection.

Soon other authors followed Fincel's success. Immediately after the first book of *Wunderzeichen* appeared, a Basel humanist, Conrad Lyscothenes, published a massive volume of *Wonder Work or God's Unfathomable Lessons* (*Wunderwerck Oder Gottes vnergründliches vorbilden*; 1557) containing thousands of stories of strange phenomena that occurred 'from the beginning of the world' to the present time.[12] Lyscothenes's book was written in Latin, but it was also translated into German by Johann Herold and enjoyed considerable popularity in Germany. Caspar Goltwurm, an influential church leader in Nassau-Weilburg, also published *Wonder Work and Wunderzeichen Book* (*Wunderwerck vnd Wunderzeichen Buch*; 1557), in which he categorizes wonders into six types – God's wonders; spiritual wonders; wonders in the sky; wonders in nature; wonders on the earth; and wonders caused by the devil – and catalogues various historical examples based on this classification.[13]

Other authors compiled books focusing on a single type of phenomenon. For example, shocked by the great comet of 1577, Lauben pastor Sigismund Suevus compiled *Comets* (*Cometen*; 1578), in which he classified historical examples of comets in seven types based on what kinds of misfortunes followed.[14] It seems that the work was well received because ten years later, the whole work was incorporated into his new book *Mirror of Human Life* (*Spiegel des menschlichen Lebens*; 1588). Georg Caesius, a pastor of Leutershausen and a well-known astrologer, also published *Chronicle or Ordered Catalogue and Description of All Comets* (*Chronick Oder ordenliche verzeichnuß vnnd beschreibung aller Cometen*; 1579), which chronologically lists all the comets in history and describes what kinds of disasters followed them.[15]

The immediate effects of the publication of these books were the intensification of people's interest in *Wunderzeichen* and the proliferation of more discourses on celestial signs. Anyone who wanted to understand the meanings of strange phenomena could find them in these books, and if they wanted to mention celestial wonders in their own writings, they could embellish their pieces with historical examples taken from these books to convince hearers that history showed that strange phenomena really foretold disasters. As we will see in Chapters 4 and 5, Christoph Irenaeus quotes a great number of stories from the works of Fincel and Lyscothenes in his own *Wunderzeichen* writings, and Irenaeus's book on comets was used as a source when Georg Caesius compiled his own comet chronicles.

Prognosticon or *Practica*, short tracts of astrological predictions published almost annually by astrologers (in many cases they were civic physicians, mathematicians and pastors), are also an important media of *Wunderzeichen* news. This genre had been one of the most popular forms of publication since the 1470s, and hundreds of sixteenth-century astrological pamphlets are still extant. These texts analyse the movements of the stars and speculate how these astrological factors affect climates on the earth and human bodies and what could happen (like plague and wars) in future years. Since they focus on the regular movements of the planets and the zodiac, technically they are different from *Wunderzeichen* works. However, the astrologers naturally showed great interest in celestial wonders like comets and discussed the astrological influences of these phenomena and the lesson people should learn from these signs. Sometimes they incorporated their analysis of *Wunderzeichen* into their *Prognosticon* to increase their appeal to readers. For example, in his *Historic Prognosticon on the Year 1568* (*Prognosticon Historicvm Auffs 1568 Jhar*; 1568), Johann Hebenstreit predicts the weather conditions of that year based on his astrological analysis. However, he quickly finishes his predictions after four pages and starts discussing seven strange celestial phenomena of 1567–8 and their meanings.[16] The contents of these tracts, often filled with moralizing lessons, biblical phrases and threatening examples from history, show a remarkable similarity to other types of *Wunderzeichen* materials.

One significant fact we must not forget is that in early modern society, the pulpit was one of the main places where people heard the latest news and learned opinions. When a mysterious sign appeared in the sky, people expected local pastors to preach about the news and express their opinions on these phenomena. Since *Wunderzeichen* were commonly regarded as signs of God's wrath, the clergy paid great attention to celestial phenomena and exploited stories of wonder to press people to repent. Even if *Wunderzeichen* were not the main point of their sermons, preachers sometimes spiced their messages with stories of wonder, especially when they wanted to draw their audience's attention and strike their conscience.

The most likely date for *Wunderzeichen* to be mentioned in preachers' sermons was the second Sunday of Advent, because in the Lutheran ecclesiastical calendar, Christ's prophecy of his return was assigned to be read during the service on this day. Since Christ mentioned strange celestial phenomena as one of the signs of the End Times, many preachers included the topic in their sermons. Some of these messages were later incorporated into postils, collections of commentary sermons on the Gospel. In addition, as intellectual leaders of the community, they wrote tracts and books about strange phenomena, often witnessed by themselves and sometimes requested by the laity.

All of the above media are publications; but considering that the majority of sixteenth-century people could not read and the number of printed materials was still limited, oral communion was still the more important media for *Wunderzeichen* news. Indeed, some accounts of wondrous phenomena printed in *Einblattdruck* and *Flugschrift* are based on hearsay. One example is Johann Wittich's *Thorough and True Report of the Dreadful and Wondrous Sign* (*Gründtliche vnd warhafftige Bericht. Von dem Erschrocklichen vnd Wunderbarlichen Zeichen*; 1561), a tract about celestial apparitions in the Mansfeld region on 27 February 1561. According to the tract, who later became a prominent physician, Wittich heard the news of the apparitions from 'a good friend' on 10 March, and as the date placed at the end of his dedication is 12 March, it seems likely that the author completed the tract within three days.[17] Wittich's account won immediate popularity, and at least two versions of his *Flugschrift* were published while his narrative was incorporated into three broadsheets. Another example is Adam Ursinus, the pastor of Tonndorf who published an account of twenty-four *Wunderzeichen* cases in 1570. Ursinus admits that the narrative of the fifth case is not based on his own observation but 'was reported to me by trustworthy people'.[18]

Of course, such hearsay was not always reliable. One of Erasmus's *Colloquies* (1524) describes how a rumour of a flying dragon in the sky (the news reported in some *Wunderzeichen* works) was started by a trickster as a joke and spread throughout the country within three days: 'it's wonderful how much the story grew in the telling. Some persons, in all seriousness, expounded the meaning of the wonder'.[19] Though the case Erasmus describes is a satirical fiction, such hoaxes probably occurred several times. According to a Strasbourg broadsheet published in 1556, a comet witnessed in Strasbourg during Easter later fell to the ground in England and destroyed several cities. Though the author of the text insists that he heard this news from 'honest people' ('warhafftige leüt'), such a disaster did not happen in Mary Tudor's land.[20] This example suggests that even in a city like Strasbourg, on an international trade route, it was difficult to obtain accurate information.

In order to enhance the credibility of the information provided, authors and printers of *Wunderzeichen* news often added the names of people who witnessed the phenomenon. Relatively reliable sources were letters reporting celestial wonders written by those who witnessed the phenomena themselves, and sometimes editors directly incorporated these letters into their *Wunderzeichen* tracts. For example, when Jacob Coler, provost of Berlin, published a short book (1595) on the wondrous incidents that occurred in Mark Brandenburg during recent years, he included an account of these celestial apparitions written by Zossen pastor Joachim Wollinus and stated that it was quoted 'as word to word'. In his narrative, Wollinus emphasized that these strange phenomena were witnessed also by 'high and noble pious people' and he listed several names.[21] Meanwhile, an Augsburg pamphlet (1583) on the unusual three-day-long eclipse of the sun observed in Rome is based on a letter written by a pastor sent to the Count of Schwarzburg – a very reliable text, because pastors and nobles were supposed to be the most trustworthy witnesses in early modern society.[22] Even if they were not written by the witnesses themselves, letters were important media to spread the news of *Wunderzeichen*. It is well known that Melanchthon forwarded information about various strange phenomena through his vast correspondence network.

The above examination of media for *Wunderzeichen* news shows that sixteenth-century people had ample opportunities to hear about strange celestial phenomena whether they were literate or illiterate. News of mysterious incidents circulated quickly, and the speed of publication shows the immense popularity of this genre. The clergy could easily obtain information about *Wunderzeichen* from broadsheets, tracts, wonder books, preachers' sermons, correspondences and daily conversations at streets and taverns. In addition, they certainly knew that they could easily draw people's attention through stories of wonder.

Astronomical and Meteorological Phenomena

In order to review the complicated world of early modern celestial wonders, Michaela Schwegler's well-ordered classification of *Wunderzeichen* types serves as a useful guide.[23] Schwegler first categorizes celestial phenomena into astronomical phenomena and meteorological phenomena. Astronomical phenomena include comets, novae and eclipses, while meteorological phenomena can be divided into three types: halo phenomena, where three suns or moons were frequently seen with strange circles and mysterious rainbows; *Fewerzeichen* (fire-signs), where it was often witnessed that the skies had turned 'fiery' and 'blood-red', and most likely these phenomena were the Northern Lights; and other *phantastische* (fantastic) phenomena, where people frequently saw fighting armies, rods, crosses, funeral processions, animals and various other objects in the sky. We are not sure what people saw in the sky in reality, but we can safely state that these phenomena are collective illusions that reflect their inner concerns. To be sure, Schwegler's classification does not cover all the celestial phenomena that appear in *Wunderzeichen* discourses. For example, she does not explain how we should categorize the frequently reported phenomenon where the sun or the moon suddenly changes to an ominous colour. Moreover, in some cases *Wunderzeichen* took place with a combination of several different types of phenomena. Nonetheless, these categories are useful in exploring what kinds of celestial wonders sixteenth-century people saw and how they interpreted them.

Among astronomical phenomena, eclipses of the sun and the moon were the most common incidents. As they believed that eclipses have astrological influences on the earth, people often recorded their observations of these phenomena in their diaries and chronicles, and several *Wunderbuch* authors such as Caesius and Goltwurm narrated noteworthy eclipses in their books. Based on Christ's prophecy in Matthew 24, some clergy insisted that the frequent occurrences of eclipses in recent days were signs of the imminent coming of the Last Judgement. However, besides astrological tracts, not many broadsheets and tracts reported eclipses as *Wunderzeichen*, probably because it was not always clear whether the eclipse which had just been observed was a natural phenomenon or an 'unnatural' one. Even the clergy admitted that eclipses were basically natural phenomena that could be predicted beforehand, although they emphasized that the unusual frequency of eclipses signified that something dreadful would occur soon.

The most common astronomical phenomena to appear in *Wunderzeichen* discourses were comets. Bright comets appeared in August 1531, March 1556, November 1577 and October 1580, but they were by no means the only comets of the century. According to Caesius's chronicle, comets were observed in Europe also in 1531, 1532, 1533, 1538, 1539, 1541, 1544, 1557, 1558, 1560 and 1569.[24] Some people claimed that they saw more comets in the sky than those listed. For example, Bartholomaeus Gernhard insisted that he saw eight additional comets flying parallel as he was observing the great comet of 1556,[25] but it is not clear what these strange flying objects were. Several *Wunderzeichen* discourses state that a comet also appeared in November 1572, but in reality this was a new star: a supernova exploded in the constellation of Cassiopeia, and a bright star suddenly became visible. This phenomenon triggered great controversy among astronomers and astrologers because it directly challenged the Aristotelian principle that there could be no change in the superlunary sphere.[26]

Early modern people were familiar with the ancient scientific theory of comet formation offered by Aristotle. In his *Meteorology*, Aristotle insists that the air that

> surrounds the earth is moist and hot because it is vaporous and contains exhalations from the earth, but that the part above this is hot and dry. For vapour is naturally moist and cold and exhalation hot and dry; and vapour is potentially like water, exhalation like fire.[27]

Since the regions of the atmosphere surrounding the earth are constantly moving, vapour and exhalations from the earth are squeezed, ignite and start burning – the origin of a comet. Since it is a fiery object, a comet can affect the climate of the earth with its heat: 'The fact that comets when frequent foreshadow wind and drought must be taken as an indication of their fiery constitution'.[28]

Many *Wunderzeichen* discourses and astrological tracts accept this theory, but they almost unanimously insist that the disasters brought by comets are not limited to storms and droughts. In his book on comets, Sigismund Suevus insists that the books of the learned and 'general experiences' tell that there are seven misfortunes which can follow comets: heat and drought; wind and storm; inflation and hunger; earthquake; 'mass death', which means plague; war and bloodshed; and the death of rulers and prominent persons.[29] Suevus then catalogues a great number of historical examples whereby these calamities befell lands after comets were seen. However, he does not explain why these specific disasters (especially from no. 4 to no. 7, which seem to have no direct causal relation with comets) occur. In his tract about *Wunderzeichen*, Nicolas Orphanus attempts to explain rationally why disasters such as wars follow comets. Heat affects people and provokes anger in them; consequently, this anger causes conflicts. This turns to war, which naturally brings the deaths of lords and nobles.[30] Orphanus also states that rulers and high nobles are of 'delicate and subtle complexion' and can easily succumb to sickness.[31] If this simple explanation sounds unconvincing, Caesius presented a more plausible image of the cycle of misfortunes that could be triggered by comets. In his comet chronicle, Caesius explains that the climatic anomaly caused by a comet (such as storms, excessive raining and harmful heat) destroys crops. Inflation and hunger naturally follow, and this could then turn to 'inner or civil war' ('Innerliche oder Bürgerliche krieg'). In addition, sickness and pestilence are brought on by poisonous air caused by the comet, and they kill not only the common people but also the rulers. Consequently, changes of *Regiment* (rulership) come after their death.[32]

Despite these differences in their views, most sixteenth-century *Wunderzeichen* discourses agreed that comets were signs of impending disaster. According to Johann Pilich, a pastor of Jüterbog who chose the comet of October 1580 as the topic of his New Year's Day sermon, common experience and history tell us that there was no comet which does not foretell 'God's great punishments', and what people should do is repent and leave their sins behind.[33] The 1572 nova was also interpreted in the same way.[34] Moreover, a number of authors, including Caesius and Suevus, interpreted comets as God's messengers sent to warn people of the imminent coming of the Last Judgement. Adam Ursinus, who labelled a nova with the ambiguous term 'comet star', insisted that because a comet was seen from Arabian lands to Jerusalem when Jesus Christ was born, the recent comets indicated that the Second Coming of Christ was near.[35]

There was a phenomenon that was observed more frequently than comets. Many sixteenth-century texts narrate that three suns suddenly appeared in the morning sky with one or more 'circles' or rainbow-like objects encircling them. Scientifically, this phenomenon is called parhelion, commonly known as a 'sun-dog' or 'mock sun', or in sixteenth-century German, *Nebensonnen*. The natural cause of parhelia is tiny ice crystals in hexagonal-plate shapes floating in the cirrus clouds in the upper troposphere. When the sun's rays pass through these plate crystals, they are refracted, and this gives human eyes the impression of sun rays in three different locations. Sometimes the 'three suns' are accompanied by one or more circular halos, while a huge arc that looks like an upside-down rainbow (called a circumzenithal arc) can appear high above the sun. Though it is rare, a 'sun pillar', a column of light standing upright in the sky, can also appear. All of these objects are phenomena created by the refraction of the sun's rays, but which object appears depends on the shape of the ice crystals and the path of the sun rays through them.[36]

The authors of *Wunderzeichen* reports carefully observed the halo phenomena and recorded how they appeared, changed their shapes and vanished. Many wondrous phenomena which Adam Ursinus recorded in his 1570 pamphlet are *Nebensonnen* phenomena. For example, on 22 December 1568, shortly after sunrise, it was seen that a long stream of light with a sharp point emanated from the sun. Thereafter, two bright *Nebensonnen*, also emanating light upward, appeared on both sides of the sun. One rainbow was seen above the true sun, and two 'rainbow-colour' columns with sharpened points stood on both sides of these three suns.[37] Ursinus attached a rough illustration that he had probably drawn himself. He also stated that in the evening of the same day, three moons appeared in the sky accompanied by 'round, long *Balcke* [balk]' in rainbow colours.[38]

What drew people's attention to this phenomenon is the common belief that this was a portent of impending misfortunes, especially on the political level. Johann Hebenstreit believed that the appearance of three suns forewarned of war, change in religious and political authorities and death of rulers.[39] Inspired by his own witnessing of the three suns in the sky over Erfurt, Hebenstreit compiled a book titled *Wunderzeichen* (1564). In this work, Hebenstreit catalogues various historical examples of the three-sun phenomena and the disasters which followed. Such phenomena were seen prior to the Peasants' War, the Schmalkaldic War, the Siege of Magdeburg, the Interim Crisis, the Marian persecutions of English Protestants, the French War of Religion and the deaths of Luther, Melanchthon, Charles V and Henri II – virtually all of the major political events in his century.[40]

Why did sixteenth-century people interpret mock suns in such a way? A Nuremberg broadsheet about *Nebensonnen* in January 1580 offers a clue. After the three suns appeared, four mysterious rainbows emerged one after the other above the suns, while a huge white cross was also seen near the rainbows. The author of the text interprets that the sun symbolizes *Regimentspersonen* and teachers of the Church, while the mock suns mean false forces creeping into the authorities to undermine legitimate power.[41] In his exposition on three suns and an upside-down rainbow seen in 1581, Wolfgang Peristerus, a preacher of Berlin, presents a similar interpretation. He states that the three suns indicate 'a split in the teaching of one Christian Church' and collapse and change of political *Regimenten* (rulership/authorities). Since Christ is compared to the sun in Malachi 4, multiplication of the sun signifies the rise of 'heretics and seducers in churches and schools'.[42] Thus since both the Church and political rulers were compared to the sun in the sixteenth century, false suns could not be a good sign. Indeed, some broadsheets assert that three suns foretell the final change of earthly regimes – the Last Judgement.[43] The common lesson of this type of *Wunderzeichen* can be summarized in what Peristerus wrote in the conclusion of his exposition: let us not ignore this warning shown by God; we should repent sincerely, always be awake spiritually, and wait for the coming of Christ.[44]

In addition to the above phenomena, people in sixteenth-century Europe were often shocked to see how the whole sky suddenly turned 'blood-red' as if it were filled with fire. This phenomenon, commonly called *Fewerzeichen* (fire-signs) or *Chasma/Chasmata*, is generally interpreted as the Northern Lights today.[45] The shock and bewilderment of the people who saw the sudden appearance of this celestial fire is described in Spangenberg's Mansfeld chronicle:

> On 28 December [1561], between five and seven of early morning, people saw a very dreadful fire-sign in the sky, between East and North. It was such a terrible sight that no people living at that time had ever seen such a thing. [They] saw nothing but the sky burning, and under the flames pure blood was running.[46]

This aurora on Holy Innocents' Day was observed in many German lands, and several broadsheets and tracts were published to report the strange incident. One of them is George Licht's tract which depicts the changes in the sky observed in Frankfurt an der Oder. According to this account, suddenly the sky opened like an eye, and some people say a black cloud appeared, while others say it was 'a black man'. It (or he) emanated white sulphur-like fire, and the fire spread first towards the southeast and then towards the northeast. Later 'the black cloud or man' poured out blood-red fire, and the flames spread in the shape of 'a long stream or stripe, like a long, thin timber or spear'.[47] Though somewhat fantastic, the description captures the wavelike movements and colour shifts of the Northern Lights.

Similar *Fewerzeichen* were recorded by many authors throughout the latter half of the sixteenth century. Nicolas Orphanus's pamphlet describes the celestial lights he witnessed in November 1574. At ten o'clock at night, the sky suddenly became bright as if the moon was shining. A black cloud like a huge mountain appeared, and three rays of clear light 'like flames of fire' emanated from the top of it. Soon 'a big fiery column' stood on the mountain-cloud and brightened the sky.[48] According to a tract on *Fewerzeichen* written by Magdeburg pastor Johann Baumgarten, 'a fiery crack (*Fewersklufft*)' was seen in the sky after sunrise, and then a 'blood-red fiery cloud' appeared. As the cloud spread over the sky, it looked as if the whole sky was filled with blood and fire. Amidst the flames, large fiery rays and '*Pyramides*, like pointed and burning towers' stood upward from the horizon.[49]

How did sixteenth-century people interpret the meaning of *Fewerzeichen*? Blood and fire reminded people of only one thing: wars. Many authors agreed that this strange sight was a portent of future wars and disorders. Hebenstreit noted that *Fewerzeichen* generally meant future war or the death of rulers, and reminded readers that similar phenomena had appeared shortly before the Turks' Siege of Vienna in 1529.[50] Fincel asserted that the meaning of celestial fire was 'always great hostility, war, and change of regimes'.[51] Baumgarten insisted that such celestial flames were warning signs of the coming of the Last Judgement because according to 2 Peter 3, the world was to be consumed by fire at the very last. However, after discussing the connection between fiery signs and the Last Judgement, he stated that these signs could also foretell God's punishments such as 'war, inflation, contagious disease and pestilence' and recounted various dreadful disasters in the sixteenth century that were preceded by fiery signs.[52]

What could people do to escape such catastrophes? As the cause of this celestial fire is God's burning wrath on account of our sins, Baumgarten admonished, we must eliminate the source of fire. Since the 'straw and wood' which inflame God's wrath are our sins, let us extinguish the fire with our tears.[53] In the words of Johann Cuno, who published a book on this phenomenon in 1578, fiery signs are a 'heavenly new preacher' who threatens the unrepentant with God's wrath and encourages believers with his grace and forgiveness.[54] Whether or not they were willing to listen to the pious admonitions of these preachers, the impressive sight of the blood-red sky must have given people a solemn feeling and made them uneasy about the future.

Celestial Apparitions

The phenomena examined above can be explained from scientific perspectives, but there are many *Wunderzeichen* reports which cannot fit into any category of natural phenomena. One example is when sixteenth-century people saw in the heavens various figures and objects, such as two armies fighting each other, as if the sky had turned into a huge film screen. Schwegler notices that many of these celestial apparitions had a red-colour element such as blood or fire and suggests that these visions may be products of the Northern Lights. According to this view, visions of two fighting armies were in fact movements of the aurora.[55] Indeed, when Johann Schütz, a Wittenberg professor of theology, describes the movement of *Fewerzeichen* in his book on *Wunderzeichen* (1582), he states that 'the sky looks fiery and blood-red, in clouds people see as if light beams are engaging in a skirmish like two armies moving together'. Schütz repeats the same comparison in a later part of the book, and this suggests the possibility that the movements of the Northern Lights induced people's illusions of fighting armies.[56] This theory can be applied to other kinds of apparitions too. According to Fincel, shortly before Elector Moritz of Saxony was killed in battle, a vision of a big man covered with wounds appeared in the sky. Blood was running from all of his wounds, and sparks appeared on his body. This image strongly reminds us of the 'black man' emanating blood-red flames described in Licht's tract. Both of the images can be interpreted as illusions created by the Northern Lights.[57] However, it is true that there are a number of apparitions which cannot be dismissed as auroras. A Strasbourg broadsheet reports that in June 1554 the people of a village near Nuremberg saw two armies of knights fighting each other with long spears. This incident, also reported by Fincel, could not be a product of the Northern Lights because it occurred during the daytime in the summer season.[58]

Instead of trying to explain these visions scientifically, it is more fruitful to consider *why* people saw such images. We can say that these apparitions were collective illusions most likely triggered by some kind of meteorological phenomena. As Alexandra Walsham points out, the illusions and interpretations added to these phenomena reflect sixteenth-century anxiety and offer unique windows into popular mentality.[59] Although it is dangerous to analyse the minds of people from a different century, based on limited amounts of sources, some speculations may offer insights.

According to the entries in Fincel's book, celestial battles were frequently seen in 1553–5. Two months after the apparition of a battle fought in the sky near Nuremberg, a similar battle was witnessed in Danzig, where two armies in black fought each other, separated, regrouped and resumed their fighting three times in the night sky.[60] Similar visions were also observed in June 1553 and August 1555.[61] During these years, Emperor Charles V and the Protestant princes were continuing negotiations to end the unstable political condition created by the disastrous Schmalkaldic War. Perhaps celestial apparitions of fighting armies can be interpreted as a reflection of the people's fear of the renewal of warfare. Fifty years later, when the Stralsund Superintendent Conrad Schlüsselburg mentions *Wunderzeichen* in his postil, he states that 'people saw armies in the sky during the time the German War was going on'. Though this kind of apparition was also seen during peacetime, Schlüsselburg's reference suggests that it was remembered as a wartime phenomenon.[62]

The apparition of a funeral procession or a bier often appears in *Wunderzeichen* literature. According to an Augsburg broadsheet (1561), some people in Emskirchen saw two biers in the sky covered with a black cloth. Several persons in black were standing beside the bier, and there appeared several grave holes near the group.[63] It seems that this news was widely circulated, as a report of this *Wunderzeichen* appears in the visitation record of Nuremberg.[64] What did the vision signify? A Strasbourg broadsheet (1574) which reports a series of apparitions offers a clue: on 22, 24 and 26 November the people of Elzach saw three types of apparitions in the sky during the night. The first one was a bier carried by four men, the next was a fiery sword, and the last was a giant rod. According to the author of the text, the funeral procession meant the death of many people ('gemeines Landssterben', usually plague) or the death of the rulers; the sword foretold a dreadful war in the near future; and the rod indicated the long-time inflation of grain prices.[65] The fact that these signs were seen within two-day intervals signified that the disaster would come within two months or two years. Thus we can regard the apparitions of funeral processions as a reflection of the people's fear of pestilence or the death of lords, which would bring political confusion to the land.

Images of swords, which also appeared in Elzach, were frequently witnessed. Fincel recounts that 'a warrior' with a battle sword appeared in the sky over Jena in June 1561.[66] According to a tract written by Johann Schütz, a hand holding a blood-red sword was seen in the sky over Bitterfeld in 1568. Schütz recounts that an angel appeared to a girl and explained to her the meaning of this vision: it warned of three divine punishments – terrible bloodshed, dreadful pestilence and severe inflation. Thus the vision of swords could signify war, but at the same time it could mean divine punishment in general.[67] Likewise, the image of a rod, which appears in a number of *Wunderzeichen* narratives, was also regarded as a general symbol of God's punishment. When a crucifix, a rod and two columns appeared in the sky over the Mansfeld region, Johann Wittich immediately interpreted the rod to mean divine punishment. The columns showed that two foreign enemies, namely the Turks and the Muscovites, would come to punish German people.[68]

Although the above examples are only a small portion of the celestial apparitions that occurred in the period, we can grasp some of their characteristics. The most important point is that many of these apparitions contain elements which can be associated with death and disasters, and the interpretations given to these visions suggest that they reflect people's fears of wars, plague and other misfortunes. These apparitions were generally interpreted as warning signs of God's wrath, pressing people to repentance, and this lesson can be applied to *Wunderzeichen* in general. Whether written by the clergy or laity, whether in popular broadsheets or sermons, they state one point: God is about to inflict various disasters on Germany, and people must repent and turn to God to avoid his wrath and receive salvation for the Day of Judgement, which will come very soon.

Before closing this section, there is one final point to consider: if these apparitions were projections of people's inner anxiety and concerns, how were such feelings visualized? Contemporary sources suggest that the contents of celestial apparitions were based on visual materials familiar to the people. According to a news pamphlet published in 1555, a vision of Christ sitting on a rainbow appeared in the sky over Freiberg. The pamphlet states that the image of Christ was '[in the way] a man usually depicts when he paints the Last Judgement'.[69] This indicates that the iconography of this vision came from the familiar image of Christ in the Judge's seat, usually depicted in church paintings and woodcuts. When Fincel recorded the incident of the image of Elector Johann Friedrich allegedly seen in the sky (this case will be discussed in Chapter 3), Fincel stated that it was a picture 'like the one painted by Lucas the painter'.[70] This also suggests that the celestial image came from the portrait (probably in woodcut) produced by Lucas Cranach. We can imagine that the proliferation of visual images through printing encouraged people to see more visions in the sky.

However, there are several apparition cases clearly deviating from the iconography of the 'official' visual culture. For example, according to an anonymous tract, a scene of the Last Judgement appeared in the sky over Forchheim in 1556. There, Christ was sitting on the throne, holding a naked sword. He was surrounded by angels with trumpets and many patriarchs. This image fits into the traditional image of the Judgement, but the apparition included a strange sight: together with devils, there were also a number of horsemen in black on the left side of Christ's throne. The mysterious horsemen then rode into a group of pitiful-looking figures, obviously damned sinners, grabbed some of them by their hair, threw them onto the horses and took them to the devil's hordes.[71] Since there is no painting of the Last Judgement containing such a bizarre scene, most likely this image came from folkloric belief. In the thirteenth century, Caesarius von Heisterbach recorded a story about devils on horseback chasing and abusing the ghost of a priest's mistress, in exactly the same way the black horsemen in the Forchheim apparition did.[72] Evidence indicates that a similar belief of punishment for damned souls survived in later centuries. More specifically, in the early sixteenth century, there was still the widespread belief that the souls of the dead, who died in 'bad ways', were snatched by demonic hordes on horseback and carried around the world.[73] If this belief was the source of the Forchheim apparition, it suggests that the celestial vision did reflect people's inner worlds and served as a place where official theology and popular belief interacted. Though it is difficult to draw solid conclusions based on such a limited number of sources, we can safely state that people projected their inner concerns and visual experiences onto the various celestial apparitions.

Climatic, Social and Political Environments of *Wunderzeichen*

The 'repent or disaster' message, endlessly repeated in discourses on wonder, should not be regarded as a mere stock phrase. Unless it related to the reality of people's everyday lives, *Wunderzeichen* messages could not acquire such a level of popularity in society. The harsh social and political conditions of German society during the latter half of the sixteenth century show that people had ample reason to fear possible misfortune and to seek God's plan for the future. The major dangers during the age can be categorized into three types: a cycle of crop failure, inflation, hunger and pestilence brought about by climatic anomalies; war; and change of *Regimenten*. All of these misfortunes are regularly invoked in *Wunderzeichen* literature.

Recent climatologists generally agree that the latter half of the sixteenth century was in the midst of a period of general cooling known as the Little Ice Age. This cooling period most likely started around 1300; however, the climatic anomalies caused by this temperature drop became more apparent after 1550. A study by Christian Pfister and Rudolf Bràzdil shows that rainy, cold summers and severe winters became the dominant climatic feature in Central Europe from 1560, and this trend culminated between 1587 and 1597. Consequently, the overall price of rye in German towns reached its peak between 1565 and 1600, while wine production dropped markedly during the last two decades of the century.[74]

The period 1569–73 can be called 'the first marked phase of "the Little Ice Age"',[75] but records of the early 1560s already show the bewilderment of people suddenly facing a change in climate. In 1562 Johann Pffefinger, a prominent theology professor of Leipzig, expressed his concern that the recent weather conditions for summer and winter were unusual and few people had seen or heard of such climatic anomalies. The weather was like people in these days, 'furious, wild, stubborn and unstable', Pfeffinger added.[76] Spangenberg recorded that in December 1564, rain, snow and fog continued for eight days and people never experienced such cold weather. Though there was a brief respite, extremely cold temperatures returned after New Year's Day and continued until March.[77] Other texts state that after a *Chasma* was seen in February and October 1564, it was followed by 'dreadful coldness, terrible pestilence, and inflation ... Men said that many people in the Netherlands region lost noses, ears, hands and foots by frostbite'.[78] Most likely this is the winter immortalized by Pieter Bruegel the Elder's painting *Return of the Hunters*.

However, as Wolfgang Behringer states, the 'first real crisis of the early modern "Little Ice Age"' came in 1570. Untimely rain brought heavy harvest failures in crops and hay, causing a serious inflation of grain prices. To make the situation worse, the blockade of Mediterranean trade by the Turkish navy and the civil wars in France and the Netherlands severely disturbed the European economy and caused mass unemployment in cities.[79] This crisis continued for several years, as Johannes Mechtel wrote in his chronicle: 'The years 1570, 1571, 1572, 1573 were all terrible and troublesome year, and in the year 1574 the heaviest inflation people's mind can imagine followed'.[80] Suffering of the poor during this period was recorded in many contemporary narratives. Again, Spangenberg's chronicle vividly conveys the voices of the ordinary people in the Mansfeld region. The year 1570 ended with such unusually heavy snow and strong wind that 'the elder people cannot recall that they saw such snowing during their lifetime'. The snow trapped many travellers and their wagons heading to the Leipzig fair, and not a few people froze to death. Searching for food, black foxes and wolves came down near the villages. After the thaw season, *Heubtkranckheit* (perhaps influenza?) spread throughout the region, although most sufferers later recovered. In the spring grain price became higher 'day by day', with 'greater burden for the poor'. Throughout the year 1571, 'dear bread ['liebe Brot'] became smaller and more expensive and the poor had difficulty to obtain them'. Then came the dreadful thunderstorm on the Day of Saint Margaret, devastating all of Mansfeld and the surrounding regions: 'It is not possible that man to write or say how terrible and dreadful the storm was ... Many thought that the Last Day was at hand, heaven and the earth would go away, and the Lord will come for Judgement'.[81]

Although it seems that the people of Mansfeld escaped the pestilence in 1571 (although they suffered the less severe *Hauptkranckheit*), usually the cycle of severe weather, inflation and epidemic in early modern communities was common. Historians point out that the plague usually followed inflation and hunger, as people's bodies, weakened by malnutrition, became easy prey for epidemical diseases.[82] An anonymous Augsburg chronicle records that 'the great *Pest* ruled Germany' during the early 1570s, and 'in some places almost all [inhabitants] of towns and villages died out', plus 'a great inflation and hunger prevailed in the land, especially in Augsburg'. Indeed, the register of the dead in Augsburg shows a dramatic increase of mortality in 1570, 1571 and 1573.[83]

This patterned cycle of climatic anomalies and social disasters during the 1570s was repeated throughout the 1580s and the 1590s, with only a few intervals. Winter was almost unfailingly more severe than average between 1586 and 1595, while cold and rainy summers continued throughout 1591–8.[84] Flooding of the major German rivers such as the Main, the Pegnitz and the Middle Elbe in the sixteenth century was disproportionally concentrated in the 1560s–1590s.[85] These climatic anomalies and natural disasters continuously affected the harvest and grain prices. Although prices were relatively cheap in 1575–8, severe inflation returned in 1579–80, and after four years of respite, inflation plagued German lands again and continued until 1594, except for two cheaper years. As expected, plague continuously raged during these difficult years.[86] It is not strange that people undergoing such a difficult time looked for scapegoats on which to blame their anger and frustration. Behringer points out that there was a correlation between the increase of witchcraft trials after the 1560s and the harvest failures brought by the Little Ice Age climate.[87] Arrests and executions of witches, who allegedly allied with the devil to destroy Christian communal lives by bringing storms and hailstorms, heightened the apocalyptic atmosphere of the age.

Considering this background, it is no wonder that sixteenth-century people showed great concern about *Wunderzeichen*, especially comets, because it was scientifically believed that a comet could affect the climate by its heat. This means that a comet could destroy the fragile balance of an agrarian society and trigger a dreadful cycle of misfortunes. The original passages by Caesius about the effect of comets, which shows his knowledge of this deadly cycle, conveys to us the tone of anxiety which sixteenth-century people felt: 'gross vngewitter

/ zu gewiseer zeit grosse kelt vn[d] vberflüssige regen / bald vbrige truckenheit vnd grosse schedliche hitz / daher verderbung der früchte oder vnfrüchtbarkeit / theurung vnd hunger erfolget / … dazu innerliche oder Bürgerliche krieg'.[88] Paraphrased, this passage means that thunderstorms, severe cold, excessive rain and severe drought bring crop failure, followed by inflation and hunger. These events also cause rebellions, which may mean bread riots and other kinds of disorder. Caesius wrote this passage at the end of the troublesome 1570s, and by experience, many readers knew how dreadful this cycle of misfortunes was.

Climatic anomalies were not the only starting point for inflation, hunger and pestilence. War could bring all of these disasters, plus the dangers of physical violence, plundering, fire and heavy taxation. Although there were few national-scale military conflicts in Germany between the Peace of Augsburg and the outbreak of the Thirty Years' War, people were constantly living under the shadow of warfare because 'infidel' armies were threatening the borders of the Empire. We have already seen that the vision of two columns that appeared in the Mansfeld region was interpreted as a warning of invasions by the Muscovites and the Turks; the danger of these two forces was repeatedly invoked in *Wunderzeichen* discourses.

In 1558 the army of Ivan IV of Russia suddenly invaded Livonia, the Baltic regions mainly ruled by the Livonian Order of the Teutonic Knights. The defence lines of the Order quickly collapsed, and the war, which lasted until 1583, ended with the partition of Livonia, because the lords of the Livonian territories, in a panic, offered their lands to the kings of Poland, Sweden and Denmark in order to gain their protection.[89] This invasion terrified many people in Germany. The Muscovites had largely been an unknown tribe for most Germans until the invasion, and the fact that the land of the German knights was destroyed by a mysterious army from the north gave an apocalyptic flavour to the news. It is natural that people recalled Ezekiel's prophecy that Gog and Magog would come from the north to punish people prior to the Last Judgement (Ezekiel 38). Reports about the dreadful atrocities committed by the Muscovites were circulated in cheap prints, and the clergy did not miss this opportunity, invoking the danger of invasion to terrify their listeners and press them to repent.[90] The spread of sensational accounts, threatening messages by preachers and the lack of reliable information all intensified the people's wild imaginations.

Some *Wunderzeichen* discourses claim that the Livonian War had been anticipated by celestial signs. Hebenstreit notes that the outbreak of the war was foretold by *Fewerzeichen*.[91] Johannes Renner, an eyewitness of the invasion who managed to survive the turmoil, opens his chronicle of the Livonian War with the appearance of the 1556 comet, whose tail stretched across Livonia and Russia.[92] Though it might be in hindsight, obviously Renner regarded the comet as a portent of the subsequent catastrophe. Thus the war offered evidence that celestial portents must not be disregarded.

In *Wunderzeichen* discourses, usually the Muscovites were mentioned along with the Turks. In his tract about the three suns and the moons observed in Erfurt in December 1568, Caspar Fagius states that these signs indicated that Germany would be attacked by forces from the north and the east, which means the Muscovites and the Turks.[93] Nicholas Selnecker, the Superintendent of Leipzig, wrote that if the people ignored God's call to repentance shown by the great comet (which appeared in 1577), and remained in their sins, 'the Turks, the Muscovites, and other enemies will ... befall our necks'.[94] Unlike the Muscovites, the Turks had been threatening Central Europe for almost a century. The danger of Turkish invasion suddenly became real when King Ludwig of Hungary died in the battle against the army of Sultan Suleiman I (the Magnificent) in 1526, and Hungary was thrown into chaos. This tragedy was followed by Suleiman's invasion of the mainland of Austria and the Siege of Vienna in 1529. Though the Turkish troops withdrew one month later, the memory of the terror traumatized many people over the next several decades.[95] The Turks constantly threatened the Empire with military advances from the 1530s to the 1560s. The invasion of the Habsburg territory of Hungary in 1566, which brought the fall of Szigetvar, especially shocked the German people. The war which broke out between Austria and the Turks in 1593 dragged on until 1606.[96] Though the mainland of Germany did not suffer from a direct invasion, people were constantly haunted by the fear of the Turks. Throughout the sixteenth century, hundreds of *Türkenbüchlein* were published to report the atrocities committed by the Turks and to impress upon the people the nightmarish danger of their advance. In Gregory Miller's words, throughout Europe 'the "Turk" became a catchword for popular fear and anxiety'.[97] People saw an apocalyptic implication in the advance of the Turkish infidel, as Fincel confidentially identified the apocalyptic tribes Gog and Magog as 'the Turks or Ottoman emperor'.[98]

It was not strange for people to see the shadow of the Turks in *Wunderzeichen*. According to an Erfurt broadsheet (1581), after the three suns were seen, two armies of knights and foot soldiers appeared in the sky and fought each other. Later a third army, which looked like 'Turks or Saracens', came from the north and surrounded the two armies. Does this vision reflect the people's fear that the power struggles within the Christendom will be exploited by the Turks?[99] The two celestial apparitions reported in 1593 seem to have been inspired directly by the outbreak of war against the Turks during the same year. According to one tract, the inhabitants of Rackersburg (Bad Radkersburg) saw a group of people marching in the moonlit sky. The banners they carried had 'a half-moon and star, like [in] the Turkish manner'. Soon an apparitional army of Christians also emerged from the cloud, and battle between the two armies continued for several hours.[100]

Those authors of *Wunderzeichen* discourses which emphasized the danger of the Turks' advance could claim that their interpretation was not a wild guess but rather was based on historical grounds. The fact that *Fewerzeichen*, the three suns and a comet were seen prior to the Turkish invasions of Hungary and Vienna had become part of common *Wunderzeichen* lore, and astrologers and preachers fuelled the anxiety of the people by quoting these past examples and applying them to new celestial phenomena. In his tract on the 1577 comet, Samuel Müller insists the comet's appearance in the Taurus signified the danger of the Turks because comets had also appeared in this location prior to the Siege of Vienna and other Turkish attacks.[101]

There is another type of danger that threatened Lutheran communities. We have already seen that Peristerus interpreted the three suns in 1581 as signs foretelling the split within the Church (brought by 'heretics and seducers') and change in the political *Regimenten*. We do not know what kinds of changes Peristerus imagined here, but the fear of 'change of rulership' is expressed in many *Wunderzeichen* discourses. Behind this phrase, we can see the ominous shadows of the religious conflicts raging in France and the Netherlands, the advance of militant Counter-Reformation Catholicism, and the rapid spread of Calvinism threatening the future of the Lutheran Church. Moreover, as we shall see in later chapters, the Lutheran Church was constantly plagued by theological controversies throughout the latter half of the sixteenth century, and theologians attacked each other as 'false teachers' and *Schwärmer* (enthusiasts, or mad).

Many Lutherans harboured deep suspicions that the papists were plotting against them. Even as early as 1561, in his booklet about *Fewerzeichen*, Georg Licht, who believed that the fiery sign foretold a large-scale war, expressed his suspicion that the 'hellish father pope' was preparing for a 'blood bath' ('blutbadt') through his minions.[102] Licht hoped that God would protect 'his [God's] Church in Germany' from 'the pope's, the Turks' and the Muscovites' Tyranny' plus the high princes' conspiracy.[103] He may have thought that his fear of war came true when civil war broke out in France the next year, but the real bloodbath took place in August 1572: the St Bartholomew's Day Massacre. Fear swept through the German Protestants that the Catholic offensive would turn on Germany next. This shock is explicitly reflected in the broadsheet published in Schweinfurt in 1573, which reported celestial apparitions witnessed by the Huguenot citizens of La Rochelle besieged by the French king's army. According to this narrative, a group of people were seen in the sky, kneeling in front of soldiers who were holding bloody lances pointed at them.[104] Although the text narrates that the people of La Rochelle were rather encouraged by this strange apparition, the woodcut of the people begging for their lives in front of the armed enemy had to have given an ominous impression to readers.

News of the atrocities committed by Spanish troops in the Netherlands also drew the German people's attention. In his tract about the 1577 comet, Johann Creat connected the comet (actually nova) of 1572 with the war in the Netherlands and invoked the memory of the 'Spanish Fury' in Antwerp on November 1572, which claimed 9,000 lives. Creat's conclusion is that the comet recently seen was the sign of impending 'inflation, war, bloodshed and other misfortunes'.[105] The fact that this was not empty rhetoric is attested by the Augsburg people's reactions to the introduction of the Gregorian calendar in early 1584. Lutheran clergy and citizens strenuously opposed to the new calendar associated this with the pope, as they regarded it as the first step to the re-Catholicization of their community. After the Catholic-dominated city council hired foreign mercenaries to maintain the security of the city, ominous rumours that the soldiers were secretly planning to kill Protestants, plunder their houses and 'have an Antwerp Martin's night [the Spanish Fury]' spread. Terrified by the rumours, some citizens even tried to protect their houses by hiring armed guards. When the council attempted to expel the Lutheran Superintendent of the city secretly for his inflammatory anti-Catholic preaching in September 1585, it provoked a riot. The danger of a 'Parisian Wedding' or an 'Antwerp Kermis' was very real in the minds of the Augsburg Lutherans.[106]

In 1584, during the midst of the *Kalenderstreit* (calendar conflict), a tract was published in Augsburg that reported a strange eclipse which had allegedly occurred in Rome. After recounting the pope's insolent attempt to introduce a new calendar, the author described how the city of Rome was covered with darkness for three days. Strangely, the sun could be seen two or three miles away from Rome.[107] The report suggests that the tyranny and blasphemy of the Antichrist-pope, symbolized in his attempt to change the dates of the holy days, met with a warning from heaven. Though the tone of the text is aggressive and sarcastic, this may reflect the frustration of the Lutherans, who failed to counter the continuing advance of Counter-Reformation influences. The Lutheran community of Augsburg had to agree to accept the new calendar after a brief respite.

The spread of religious controversies within the Lutheran Church caused serious disunity among believers, and many people could not hide their concern that the Church was unable to stand up to the papists and any other enemies under such conditions. Since we shall see many examples of concerns about 'false teaching' in the clergy's *Wunderzeichen* literature in later chapters, here we will examine only one voice. In his brief exposition on the 1577 comet, Simon Pauli, a prominent theologian in Rostock, stated that 'without doubt' this new 'false star' (the comet) forewarned not only of war, bloodshed and plague but also of the rise of heresy and other misfortunes for the Church. Pauli noticed that the day when the comet appeared, 10 November, was the ninety-fourth anniversary of Luther's birth, and he interpreted this 'false star' as a sign that a 'new, false, erroneous teaching' would appear through false teachers and preachers.[108] Though he did not specify further, Pauli's view reflects the Lutherans' deep concerns about theological errors.

The most formidable 'false teachers' for the Lutherans were the Calvinists. They abhorred the Calvinists' denial of the physical presence of Christ in the eucharistic elements, and the doctrine of predestination was viewed as an abominable distortion of the Gospel. The fall of the crypto-Calvinist theologians in Wittenberg in 1574 reveals the danger of the Calvinistic influence crept into the Lutheran Church. Lutheran preachers vehemently denounced Calvinist teachers as the devil's agents. In his book published in 1580, Johann Schütz insisted that the devil is 'using primarily the Sacramentarians to deprive us cunningly and maliciously of God's saving Word, thereby moving Christ even further away from us than the old church had ever done'. Schütz asserted that Satan was using at least thirty-seven assistant devils to promote the Calvinists' erroneous doctrines. When he published a revised edition of the book eleven years later, he increased the number to fifty, probably reflecting his increasing concern and frustration.[109] In 1595 Ambrosius Taurer, a pastor of Wettin, published his thoughts on recent wondrous incidents, including various *Fewerzeichen*. As he talks about the people's negligence of God's Word, he starts thundering about the 'Calvinist wolf', 'Calvinist spirit of error' and 'false prophets' and calls them a bride of devil, beautifully clothed and pretending to be the angel of the Light.

> Pray earnestly that God supports us with his Word and controls the pope, the Turks, Calvinists and other heretic's deception and murder. And Christ the Lord, for it is getting toward evening, and the day is now nearly over. Stay with us.[110]

The last two lines are taken from Luke 24:29, and Taurer uses the words 'getting toward evening' in the sense that the world is facing its end.

Despite the Lutheran teachers' fierce attacks, Calvinism kept expanding its influence in the Empire through the conversions of princes, court advisors and urban magistrates during this period. Although the Peace of Augsburg did not recognize Calvinism as a legitimate religion, the principle *cuius regio, eius religio* offered a quick way to make crypto-Calvinism an official religion of the state. The Palatinate officially became a Calvinist state by dynastic succession in 1583, and the princes of Anhalt gradually changed their Church order into the Calvinist style in the 1590s. Electoral Saxony also experienced a short phase of a crypto-Calvinist regime under Elector Christian I between 1586 and 1591, and the visitations conducted after the death of Christian revealed that some crypto-Calvinist clergy did not hesitate to use intimidation, imprisonment and even physical force in order to impose Calvinistic reforms on local people, claiming their connections with the Elector's government.[111] These rulers' efforts to enforce pro-Calvinist liturgical changes often provoked fierce opposition from ordinary citizens, and in some cases their protests turned to riots.[112]

Most likely this was one of the reasons why 'death of rulers/change of rulership' provoked such concern in *Wunderzeichen* literature. The death of a prince could change the religion of a state in one night and decidedly transform the power balance between Lutherans and Calvinists in the Empire. For orthodox Lutherans, this was nothing but the devil's attack on God's true Church. Thus the 'change of *Regimenten*' did not mean a simple change of government. With the apocalyptic world view in the background (symbolized by the word 'evening' in Taurer's passage), it meant the ascendancy of the devil's false faith and the persecution of true believers.

Conclusion

Sixteenth-century people showed a great deal of interest in celestial *Wunderzeichen*, and they had a variety of media to obtain the latest information about these strange celestial phenomena. There were many types of *Wunderzeichen* including comets, halo phenomena, *Fewerzeichen* and celestial apparitions, but all interpretations contemporary authors attached to these phenomena can be summarized in one phrase: 'repent or disaster!' They were regarded as signs of God's wrath and impending misfortune (or the Last Judgement), and the only proper reaction for Christians was to repent for their sins and improve their moral lives.

The climatic, social and political conditions of sixteenth-century German society show that people had enough reasons to worry about future misfortunes. Climatic anomalies caused by general cooling often brought about crop failures, which in turn triggered inflation, hunger and pestilence. People were also haunted by the possibility of invasion by the Turks and the Muscovites, who were regarded as the apocalyptic tribes of Gog and Magog. Religious conflicts in neighbouring territories, the advance of the Counter-Reformation, the expansion of Calvinist influences in Germany and the doctrinal disunity within the Lutheran Church made many pious Lutherans worry about the future of their religion. Surrounded by these dangers, it is no wonder that they were sensitive to *Wunderzeichen* because these were commonly regarded as signs of divine punishment. The fear and concerns of the people were projected onto images of celestial wonders, also influencing people's interpretation of what these strange phenomena signified. In this aspect, *Wunderzeichen* prints serve as unique sources for the study of the history of emotions.

Considering this background, it is understandable that clerics were interested in celestial phenomena and tried to use stories of wonder for their own purposes. They knew about the enormous popularity of *Wunderzeichen* literature among people, and they had ample opportunity to amass knowledge about strange phenomena through wonder books, broadsheets and other literature. Certainly the clergy recognized that stories of wonder could draw people's attention easily and strike their conscience by terrifying them with news of the latest signs of divine wrath and the history of disasters that followed these celestial wonders.

2 LUTHERAN CLERGY AND *WUNDERZEICHEN* DISCOURSES

In the previous chapter, we have already encountered many examples of preachers' references to celestial wonders, which show that they shared the people's interest in *Wunderzeichen* and tried to comprehend the meanings of the mysterious phenomena. The two questions I want to pursue here are: besides their own curiosity, why did the preachers show such strong interest in *Wunderzeichen*? For what purposes did the clerics use *Wunderzeichen* in their writings?

Among the clergy, who were interested in celestial phenomena? A complete list of all the Lutheran clergy who discussed *Wunderzeichen* in their writings would include almost all the prominent theologians in the sixteenth century, and a number of them published books and tracts about this topic for popular consumption. The list of leading theologians who revealed an interest in *Wunderzeichen* in their writings includes the following names: Martin Luther; Philip Melanchthon; Matthias Flacius Illyricus; Nicholaus Medler (Superintendent of Braunschweig); Jakob Andreae (provost of Tübingen University and the chief architect of the Formula of Concord); Martin Chemnitz (Superintendent of Braunschweig and one of the co-drafters of the Concord); Andreas Musculus (professor of theology at Frankfurt an der Oder, who helped in the formulation of the Concord); David Chytraeus (professor of theology at Rostock, one of the formulators of the Concord); Georg Major (professor of theology at Wittenberg); Paul Eber (professor of theology at Wittenberg and General Superintendent of Saxony); Johann Pfeffinger (professor of theology at Leipzig); Johann Wigand (professor of theology at Jena); Erasmus Sarcerius (later Superintendent of Mansfeld); Cyriakus Spangenberg (deacon of Mansfeld); Hieronymous Mencel (Superintendent of Eisleben); Christoph Irenaeus (once a court preacher of Weimar); Basilius Faber (rector of Quedlinburg and an influential educator); Jakob Heerbrand (professor of theology at Tübingen, who taught theology to Kepler); Simon Gedike (court preacher of Brandenburg); and Simon Pauli (professor of theology at Rostock and Superintendent of the city). Among these theologians, Flacius, Medler, Chytraeus, Selnecker, Eber, Pfeffinger, Irenaeus and Heerbrand published books or pamphlets that focus on *Wunderzeichen.*

Many names of less well-known theologians can be added to the list. Johann Schütz (professor of theology at Wittenberg), Hieronymus Weller (Luther's pupil who taught theology in Freiberg), Andreas Celichius (Superintendent of Mecklenburg), Hieronymus Opitz (Superintendent of Bischofswerda), Jacob Coler (provost of Berlin), Michael Bapst (pastor in Mohorn), Petrus Victorius (pastor in Havelberg), Nicholaus Erben (pastor in Erfurt), Ambrosius Taurer (pastor in Wettin) and Adelarius Praetorius (pastor in Erfurt) all wrote books on comets and various other kinds of strange celestial phenomena. The above list shows that many authors of *Wunderzeichen* discourses occupied prestigious positions within the clergy, such as professors of theology, superintendents and court preachers; this fact suggests that *Wunderzeichen* discourses belonged to the very mainstream of Lutheran ecclesiastical culture in the latter half of the sixteenth century.

But why were these prominent clergy attracted by *Wunderzeichen*? As we saw in Chapter 1, since it was commonly believed that *Wunderzeichen* were signs of approaching danger and disasters, it was natural for the clerics, as spiritual and intellectual leaders of the community, to be concerned about the significance of wonders. However, curiosity about future events was not the only reason why they were drawn to celestial phenomena. Various theological, pedagogical, social and political factors of sixteenth-century Lutheran society are intertwined in promoting clerical interest in *Wunderzeichen*.

In order to deconstruct this kaleidoscope-like world of the Lutheran clergy's *Wunderzeichen* discourses, first, three factors that stimulated the clergy's interest in celestial wonders will be examined. They are academic (Melanchthon's promotion of astrology and astronomy); spiritual (Christ's prophecy on the celestial signs prior to the End of Days); and emotional (the clergy's own experiences of celestial wonders). We shall then analyse what general functions *Wunderzeichen* stories fulfilled in the clerical messages. They can be divided into three types: to support attacks on people's sins; to terrify popular conscience through *Gesetzpredigt* (preaching of the Law); and to reassert the authority of clergymen as the 'servants of God's Word'.

Underlying Factors of the Clergy's Interest in *Wunderzeichen*

One of the important factors that turned many Lutheran eyes to celestial phenomena was the academic influence of Philip Melanchthon, the 'German Preceptor'. Melanchthon was greatly interested in the movements of the stars and *Wunderzeichen* in general, and no matter how much he was chided by Luther, he did not hide his enthusiasm about astrology and astronomy and promoted them in the university education system.[1] His passion for astrology is well summarized in his eloquent speech delivered at the graduation ceremony of Wittenberg University in 1535. In this speech, Melanchthon emphasized that astrology is a part of natural philosophy and that we can learn about God's order and providence through observing the movements of the stars. He pointed out the importance of 'signs' in heaven and admonished people not to ignore them because they were warnings sent by a merciful God:

> For if these signs are not meant to be considered, why are they written and painted on the sky by divine providence? Since God has engraved these marks in the sky in order to announce great upheavals for the states, it is impiety to turn one's mind away from their observation. What are eclipses, conjunctions, portents, meteors or comets if not oracles of God which threaten great calamities and changes for the life of men?[2]

For Melanchthon, the ultimate purpose of observing the movements of the universe was not to predict future events; the most important goal was to grasp the divine order of the universe and to understand God's providence working in this world: 'God desires that knowledge of these wonderful courses and powers should lead us towards knowledge of the divine'.[3] Thus students of Wittenberg were encouraged to turn their attention to the sky to seek out God's will.

Melancthon's sophisticated arguments fascinated many students and implanted in their hearts an interest in celestial matters. For example, Martin Chemnitz was introduced to astrology when he was studying at Wittenberg in the 1540s. In his autobiographical writing, he states that 'I afterwards pursued [astrology] with so much delight that I might have become an expert in that territory, if I had not later on turned to Theology'.[4] Though not all students shared Melanchthon's fascination with astrology, it did not mean that they were indifferent towards *Wunderzeichen*. For example, Christoph Irenaeus, a pupil of Melanchthon, rejected astrology as idolatry, despite the fact that he was an enthusiastic collector of stories of wonder.

It is well known that Melanchthon often exchanged news of *Wunderzeichen* with his friends. 'A battle in the sky over Hungary' and 'a fiery ship witnessed in the sky over Breslau' are examples of wonder which he reported in his letters.[5] When he saw a comet for the first time in his life in 1531, Melanchthon sent many letters to his friends, including Joachim Camerarius and Johann Agricola, and discussed what the comet signified.[6] When Camerarius published the first wonder book in Germany in 1532, Melanchthon warmly contributed a preface.[7] Indeed, it seems that the Preceptor even wrote at least one text of *Wunderzeichen* news for a broadsheet published by a Wittenberg printer in 1551.

It is not strange that Melanchthon also shared *Wunderzeichen* stories with the students in his class. According to one report, during a lecture in 1559 he told his students that 'soon a dreadful, sad time will come' because strange signs were appearing every day. As an example, he described the story of celestial apparitions witnessed by some Braunschweig merchants about ten years ago and explained the possible meanings the phenomena indicated.[8] Such stories of wonder told by their Preceptor had to have stimulated the young students' intellectual curiosity. It is interesting that both Fincel and Goltwurm, the two pioneer authors of wonder books, had Wittenberg backgrounds.[9] Was their interest in *Wunderzeichen* inspired by Melanchthon's lectures?

Since Wittenberg was the centre of education in Lutheran Germany, the number of students influenced by Melanchthon was enormous. In addition, his humanist curriculum became a model for other Protestant universities, gymnasia and newly created institutions. Thus Melanchthon's emphasis on natural philosophy, including astrology and astronomy, had a strong influence in the world well beyond Saxony, and historians of science recognize that Melanchthon's ideas on natural philosophy greatly contributed to the development of astronomy in early modern Germany.[10] Certainly his passionate interest in astronomy stimulated scholars' attention to irregular movements in the sky.

However, no matter how strong Melanchthon's influence was, it would not have won such popularity among the clergy if *Wunderzeichen* had lacked the support of God's Word. Scripture mentions strange signs in the sky in several parts. The passage 'I will display wonders [the term *Wunderzeichen* was used] in the sky and on the earth, Blood, fire, and columns of smoke. The sun will be turned into darkness and the moon into blood' (Joel 2:31) was often mentioned as proof that it was God who showed people mysterious celestial phenomena. However, the passage most frequently quoted in *Wunderzeichen* discourses is Christ's prophecy about the signs of the coming of the Last Judgement, described in Luke 21 and Matthew 24. Here Christ explicitly states that when the time of his return approaches, 'there will be signs in sun and moon and stars' and 'the powers of the heavens will be shaken' (Luke 21:25–6); 'The sun will be darkened, and the moon will not give its light, and the stars will fall from the sky' (Matthew 24:29). The Lutheran clergy firmly believed that the Day of Judgement was quickly approaching. For them, the revelation of the identity of the Antichrist (the papacy) and the beginning of the final struggle between Christ's Church and the devil's, materialized by persecutions of true believers, showed that the apocalyptic prophecy of the Book of Revelation was coming true. As we saw in Chapter 1, the latter half of the sixteenth century featured the deterioration of the climatic condition, the increase of natural disasters, the danger of foreign invasions and political turmoil, and these gloomy conditions surrounding Germany further strengthened the clergy's eschatological world view. The appearances of strange signs in heaven, clearly foretold by Christ, added fuel to their conviction that the Day of Judgement was at hand. In Ambrosius Taurer's words, 'because the evening of the world is coming', a righteous God admonishes people to repent not only through his 'final' pastors and teachers but also through many wondrous signs.[11]

Indeed, in the sixteenth century all Lutherans were reminded of this prophecy of Christ every December, because in the church lectionary, Luke 21 was designated for the reading on the second Sunday of Advent, and pastors were expected to preach on the topic on that day. Since Advent is completed by the coming of Christ, it was proper to contemplate the return of Christ on the second Advent Sunday. Thus every Advent season the clergy and the laity were encouraged to meditate on the prophecy of the end. In order to understand how closely celestial wonders and Christ's prophecy were connected in the clerical mentality, it is useful to examine the Advent section in sixteenth-century Lutheran theologians' postils. Postils generally refer to collections of commentary sermons affixed to the texts of the Gospels and the Epistles, and they were widely used as study guides and references for both clergy and pious laymen.[12] A number of postils discuss celestial signs and their relation to the Last Judgement in the section of the second Advent Sunday.

The theologian who set a model for the exposition of Luke 21 is Martin Luther. Though he did not put much trust in astrology, Luther occasionally expressed his interest in celestial wonders, as his remarks in the *Table Talk* indicate.[13] However, the best source for examining Luther's attitude towards *Wunderzeichen* is his sermon published in Wittenberg in 1522. In this sermon, later placed in the section of the second Sunday of Advent in *Church Postil*, he explains the meaning of celestial signs as described by Christ.[14] Luther interprets the word 'signs in the sun and the moon' as eclipses and states that they are 'tokens of his [God's] wrath and predict sure disaster for future'. Ominously, 'it seems to me, the last few years more and more frequent eclipses of the sun than in any other like period of time'. It is not a matter of whether they are natural phenomena as astronomers insist; 'though the heavenly bodies wander in their courses according to law, God has still made these to signs or token of his wrath'.[15] Other strange phenomena in the sky can also be regarded as signs sent by God:

> We have lately also seen so many comets and so many calamities have fallen from the skies and there has arisen the hitherto unknown disease, syphilis. Also how many signs and wonders have been seen in the heavens, as suns, moons, stars, rainbows, and many other strange sights.[16]

Luther mentions syphilis here probably because some believed that the illness was caused by planetary influences.

In the brief sermon in his later work *House Postil*, Luther discusses this topic again. There he teaches that within these twenty years people saw many strange things, such as 'a rainbow encircling the sun; the sun fragmented, with multiple suns appearing; also recently, within the space of two years, three comets appeared'. This means that '[c]ertainly we have warning enough just in the sun that we should not be complacent but expect the blessed return of our Lord Christ'.[17] Both *Church Postil* and *House Postil* became the most popular study texts and devotional readings among Lutheran clergy and laity, and it is not difficult to imagine that Luther's expositions became the standard of Lutherans' response to celestial phenomena.[18] Actually, Luther's followers were not bound by their teacher's interpretations. They freely amplified their interpretations on wonder with contemporary news of *Wunderzeichen* and included various strange phenomena that did not appear in Luke 21.

In his *Postilla* (1569), Johann Wigand first reminds readers that God always sends warning signs before he punishes people in order to give them a chance to repent.[19] For example, a comet was seen and strange voices were heard prior to the destruction of Jerusalem. Then Wigand turns to the eschatological signs mentioned in the Gospel. Although eclipses are natural phenomena, if they occur so many times and are 'unnatural' (meaning they appear unexpectedly), they are special signs of the End Times. As an example, Wigand mentions the strange incident of the sun turning blood-red when Elector Johan Friedrich was captured in 1547. Men must admit, he states, that there was no other age in history where so many strange phenomena were seen in the sky as in these thirty years. People often see 'battalions or armies, shooting stars, splits in the sky ... comets, fire, similar phenomena and burning elements, fiery sky'; these are the signs and *vorbild* of the Last Judgement.[20]

The postil of Rostock Superintendent Simon Pauli (1572) also shows his interest in eclipses of the sun and the moon, which 'always mean something special of great misfortune'. Interestingly, Pauli mentions several meteorological topics that may be related with the Little Ice Age climate. For example, he states that elder people claim that compared with several years ago, the sun does not have strong power both in light and in its effects on creatures on the earth. Pauli also compares the present condition of the sun with the eyes of an old man, losing power day by day and becoming darker. Facing the Last Judgement, everything is ageing, and now the sun is shining less brightly and losing its power to warm the earth.[21] Dreadful storms on the sea are also increasing, as old sailors claim. There is excessive raining, which causes damage, and actually it is the tears of heaven shed for the godless, whose damnation is at hand. In the sky there also appear many comets and burning lights named *Irrwische oder Lügenliechter*, which means will-o'-the-wisp. In these eight years the sky was seen to be filled with fiery beams that look like lances and swords. The whole of heaven burns like a fiery oven, as forewarned in Malachi 4:1: 'For behold, the day is coming, burning like a furnace; and every evildoer will be chaff'.[22] The report of the three suns that appeared in the sky over Wittenberg in 1566, witnessed by Johann Bugenhagen, is also included in Pauli's exposition.[23]

Hieronymus Mencel's list of *Wunderzeichen* (1596) contains eclipses, unnatural rainbows, multiple suns, 'amazing crosses, light beams or other images' in the sky and 'fiery signs with sparking light beams'. This last phenomenon, which appeared many times within a short period, is 'a strong reminder' of the great fire that will burn the world at the end. While experience tells that comets bring great disasters such as plagues, wars and the death of princes, they are also signs of the coming of the Last Judgement. Within three years, Mencel continues, people saw two comets whose tails were 'in the shape of a broom and a bound rod', and with these comets we are reminded that God made a rod to punish the unrepentant.[24]

Simon Gedike's postil (1588) also argues that eclipses of the sun and the moon, comets and *Chasmata* (fiery skies) were witnessed 'so frequently and generally never before'.[25] As *Fewerzeichen* and comets have become so common that they no longer draw attention, now there appeared a new star of which no man could explain the origin. The wonder-star ('Wunderstern') of 1572, which 'made all astronomers turned to be fools', remained in the sky for a while as a 'fiery rod'.[26] Georg Walther (1579) was another theologian who was attracted by the 1572 nova, stating that 'without doubt, it [the new star] signifies something important'.[27] Showing his considerable knowledge of contemporary news of *Wunderzeichen*, Walther lists cases of the sun's multiplication which occurred in 1520, 1549, 1550 and 1551. 'It is clear', he concludes, 'that [with these phenomena] God signified that the end of the world is at hand and Judgement will come to humankind soon'.[28] Comets in the shape of rods, which people often saw, meant 'God the Father threatens bad children on the earth with his wrath and punishment'.[29] Wondrous visions such as 'armies, spears, men and crosses' appeared in the sky, as had happened at Wittenberg in 1553 and 1554.[30]

The postils of Christoph Vischer,[31] Siegfried Saccus,[32] Martin Mirus,[33] Martin Chemnitz,[34] Adigius Hunnius[35] and Johann Spangenberg (for children)[36] contain similar expositions on apocalyptic celestial signs. Not all authors expressed interest in *Wunderzeichen*,[37] but the fact that these discourses about *Wunderzeichen* were included in texts such as postils suggests that they were a part of the pastors' daily study and devotion. Certainly these readings had to strengthen their interest in celestial signs, and it is natural that when they saw unfamiliar phenomena in the sky, the passage of Luke 21 was immediately recalled.

When they discuss *Wunderzeichen* in their writings, pastors use a tone similar to that used by these authors of postils. In his sermon calling for repentance (1583), Conrad Schomer, the rector of Lemgo, insisted that man has to admit that within these four or five hundred years there had never appeared so many *Wunderzeichen* as now, and 'without doubt this shows us that the Lord's Day and his Justice are not far from us; [it is also] his harbinger of a dreadful disaster and many shocking sufferings which Germany undergoes before the world's end'.[38] While listing various phenomena, Schomer asks his readers: 'How many dreadful eclipses of the sun were seen in short years? In 1563, there were five eclipses within one year. How often did two or three suns appear at the same time? How often did the sun turn to blood-red and sink?' Comets, fiery torches and rods also appeared in the sky:

> How often did strange, wondrous rainbows and other signs appear in the sky? People stood in great anxiety, [fearing] that the Last Judgement had come. How often blood-red crosses were seen in the sky? ... In many places in Germany men often saw in the sky great battles with great tumult and cries.[39]

In his well-read book about the Last Judgement (1565), Basilius Faber states: 'In recent years so many *Wunderzeichen* were seen and still appear daily' that people no longer pay attention to them. He mentions *Fewrzeichen* in the night of the Holy Innocents' Day in 1561 and states that similar phenomena were seen five times in Quedlinburg, where he served as a rector; there may even be many more signs that people have missed while they were drunk or sleeping. Faber asserts that there appeared more strange signs in these forty years ('since Luther's time') than in the past one or two hundred years, and no one can doubt that the Last Judgement is coming soon.[40]

These examples are only a small portion of the *Wunderzeichen* discourses by the Lutheran clergy, but they are sufficient to show that for these clerics, wonders in the sky and the coming of the Last Judgement were inseparably connected. It seems that many of these preachers were especially impressed by *Fewerzeichen*, which were often regarded as a new type of celestial wonders. Spangenberg once called them 'a new sign, unknown before', while Wolfgang Peristerus stated that people had not seen things like *Chasmata* for many years.[41] Johann Schütz claimed that the phenomenon was seen many times within these twenty years, which means 1562–82.[42] One of the reasons why it drew the clergy's attention might be the fact that the image of celestial fire could be connected with Peter's prophecy about the fiery end of the world (2 Peter 3). It also fits into the Scripture passage on *Wunderzeichen* in Joel 2 ('blood, fire and columns of smoke'), as Schütz pointed out. The nova also fascinated several preachers because it was believed that 'since the beginning of the world' such a thing had never appeared. According to Musculus, since Christ's first coming was signified by a new star, 'without all doubt this one [the 1572 star] signifies and proclaims the Second Coming of the Lord'.[43] Thus virtually all kinds of unusual phenomena in the sky were interpreted as signs of the End Times.

Lutheran theologians shared the impression that the number of strange phenomena was increasing in their days. Phrases like 'experiences tell us that the strange signs are increasing' often appear in their discourses. Probably their impression was true – not because strange phenomena appeared more frequently than before, but because the output of *Wunderzeichen* reports rapidly increased as the number of printers in Germany grew after 1550. However, there is another important reason why the clergy held such an impression. Faber's statement that these signs had been increasing especially 'since Luther's time' ('von D. Luthers zeiten') gives us a clue. For Lutherans, Luther's revelation of the Antichrist was the beginning of the apocalyptic era; the final countdown towards the Last Judgement had started with Luther's protest, and it was not strange that the number of wondrous signs increased as the Day of Judgement became near. Obviously Hiob Fincel deliberately set the starting point of his chronicles of *Wunderzeichen* from the year 1517, and throughout his three volumes he preserved this basic timeframe. For Fincel and other Lutherans, this timeframe was set up by God according to his apocalyptic plan. Thus celestial phenomena, the End Times and Luther's Reformation were intertwined in the Lutheran world view.

One important factor which strengthened the clergy's interest in celestial wonders was their own experiences. Many clergy had a chance to see the unusual phenomena in the sky for themselves, such as the great comets in 1531, 1556, 1577 and 1580, as well as the 1572 nova, which remained in the sky for a considerable time. The *Fewerzeichen* on Holy Innocents' Day of 1561 is mentioned in a number of sources obviously because the phenomenon was witnessed in many places of Germany. Such an experience had to leave a strong impression on people's minds. In his sermon about the 1577 comet, Heerbrand recalls the comet he saw in the summer fifty years earlier and states he had it 'still in good memory' ('noch in gutter gedachtnuss').[44] But for the clergy, it was more than just impressive. In celestial *Wunderzeichen*, they saw Christ's Words literally come true and God's plan of salvation become visible.

Such experiences gave the clergy a strong motivation to write sermons or books urging people to reflect on the meaning of these signs. In 1567 three suns and two rainbows suddenly appeared in the sky over Frankfurt an der Oder, and many who were on the way home after worship service witnessed the sight. From one edge of the rainbow a 'very long white beam' came out towards the direction of 'the Land of the Muscovites and the Tatars'.[45] Andreas Musculus, a pastor and professor of the University of Frankfurt an der Oder, obviously saw the phenomenon himself, and the sight reminded him of the biblical prophecy of Gog and Magog. Worrying that this phenomenon was the warning sign of a Muscovite invasion in the near future, he compiled a book on Luther's prophetic remarks, which urged the people to repent to avoid foreign domination. A similar case is Christoph Irenaeus, a pastor of Eisleben, whose first *Wunderzeichen* book was directly inspired by his experience of strange celestial phenomena in February 1564: after midnight, some fiery clouds appeared in the sky over the city, and Irenaeus rushed to the marketplace of the city to see what was going on; but finding that he would get a better view higher up, he returned to his church and climbed up the bell tower with other church staff.[46] We will discuss the details of his experience in Chapter 4, but the fearful vision he saw from the tower became the starting point of his lifelong career as a *Wunderzeichen*-story hunter.

Several other authors wrote down experiences of celestial wonder in their works. Hieronymus Opitz published a pamphlet on *Fewerzeichen* of the Holy Innocents' Day in 1561 and recorded how people of Bischofswerda reacted to this fiery sight. According to the author, the sky suddenly became fiery red at 6 o'clock, and everyone believed that the barns outside the city or in nearby villages were on fire. The fire bell was rung, and people rushed with buckets to form a fire-fighting brigade. As he heard the cry of 'fire!' in his study, Opitz looked outside and immediately thought it was a real fire. Running outside, he soon found that it was a not an actual fire but a 'heavenly fiery sign' ('himlich fewrigzeichen') shining. Opitz's vivid narrative shows that *Wunderzeichen* were really shocking for people and could cause panic in the community.[47]

Bartholomew Gernhard, who later became a court preacher of Weimar, also witnessed a strange sight in April 1556 and recorded his experience in the dedication of his sermon collection. In March of that year a bright comet appeared and remained in the sky for some time. The next month Gernhard, a reader of Fincel's *Wunderzeichen* volume, woke up at two o'clock in the morning and went to the window to see the same comet in the sky. However, to his astonishment, he saw not only the familiar shape of the comet but also eight other comets moving north. The shapes of the new comets looked like long spears. Immediately Gernhard woke his servants and told them to fetch the sexton and the schoolteacher. The next morning he told the church staff to pay attention to celestial signs in the future and admonished them that these phenomena had to be divine warnings.[48]

In September 1580 Jacob Coler witnessed strange fiery clouds and a vision of a double-headed eagle in the sky over Berlin. In the booklet about this *Wunderzeichen*, he carefully records the appearance of this vision, but he admits that his view was unfortunately limited. Coler saw the apparition 'between houses' in the city of Berlin and states that 'other people who were standing in towers or in the fields might have seen more'.[49] These words suggest that he made the best effort he could to record the vision. Peristerus witnessed the same vision of an eagle in the sky over Schwerin, and he describes the movements of the vision in his sermon.[50]

Michael Bapst, pastor of Mohorn, published a booklet on the *Fewerzeichen* which appeared in March 1582. There he narrates the movements of the northern light and states that this description is as 'I observed diligently and looked with heavy heart'.[51] Meanwhile, as the many entries about celestial wonders in his *Mansfeld Chronicle* indicate, Spangenberg also showed great interest in *Wunderzeichen*. The lengthy descriptions of *Fewerzeichen* in his chronicle show that, like Bapst, Spangenberg meticulously observed the movements of the mysterious phenomena. His carefulness is well attested by his reaction to the apparition of the two smoking columns which allegedly appeared in the sky between Eisleben and Mansfeld in 1561 (this incident is mentioned in Chapter 1). After narrating the rumour of the apparition, he states that he asked the miners in his town whether they saw such a phenomenon, but they replied that all of them were in the mine at that time and no one witnessed such a thing. Judging from his overall tone, Spangenberg seemed to be sceptical about the authenticity of the report of this mysterious apparition.[52]

Rural pastors also showed enthusiasm in recording their experiences of *Wunderzeichen*. In 1560 Johann Gölitz, a village pastor of Aschre, published an account of the strange apparition of a dragon he saw in the sky.[53] Thomas Wirsing, a pastor of Sinnbronn near Dinkelsbühl, also recorded in his diary *Wunderzeichen* that he himself witnessed.[54] Unlike Gölitz and other authors, Wirsing never published his observations, but he likely used them in his sermons.

These examples of the clergy's reactions to *Wunderzeichen* show that they (and the people around them) were genuinely excited by the apparition of strange signs. Gernhard did not hesitate to wake up the church staff at two o'clock in the morning to show them the strange sight, and Irenaeus and his church's staff ran up the long, dark stairs of the bell tower in order to get a better view. They also diligently recorded details of the phenomena they had seen as accurately as possible. Many of them carefully observed where the strange objects came from, in which direction they moved and what kind of shape of they had. These preachers were sure that they were seeing extraordinary works of God that were well worth recording. Musculus's interpretation of the 1567 phenomena suggests that they often tried to 'read' God's intention in the movement and shape of *Wunderzeichen*. Spangenberg's effort to ascertain the authenticity of the apparition may reflect his serious conviction that if the apparition were true, he as a preacher had to comprehend the meaning of the signs, and to convey to his parishioners God's will as revealed through the phenomenon. These experiences had to be alarming for the clergy and strengthened their conviction that God was giving serious warnings to the world. In addition, through broadsheets, wonder books and other publications, they knew well that similar phenomena were also appearing in other places. It is no wonder that they showed their anger and frustrations when they found that people were not paying enough attention to these awful signs.

These are the three major factors that stimulated the clergy's interest in *Wunderzeichen*, but the most important one is Christ's prophecy of the End Times, as apocalyptic world views were the backbone of all *Wunderzeichen* discourses by the Lutheran clergy. Again and again they reminded people that strange celestial sights were signs of the approaching of the Last Judgement, and they pressed them to prepare for Christ's return with repentance. Their own experiences of *Wunderzeichen* gave them a sense of urgency, and Melanchthon's teachings provided them with an intellectual foundation for their interest in celestial phenomena.

Calling for Repentance and *Gesetzpredigt*

Why did the clergy incorporate reports of celestial wonders into their messages? In this section, we focus on three major functions of *Wunderzeichen* lurking behind the familiar clerical mantra of 'repent or disaster!', and we examine the pastoral roles which *Wunderzeichen* discourses are supposed to fulfil. The first function is to support the clergy's attack on people's immoral behaviour and lukewarm attitude, summarized in the term *Sicherheit* (security; in this sense, meaning a false sense of security) and the neglect of God's Word. The second function is to terrify people's conscience through the 'preaching of the Law' and to lead them to the true understanding of the Gospel. Since these first two topics are interconnected, they will be analysed in one single section. The third function is to reassert the clergy's authority as servants of God, and this will be discussed in the next section.

Protestant reformers initially had high hopes that if the Gospel were taught correctly, common people would change their lives and became true, pious Christians. However, their hope was soon betrayed. The people's adoption of their clergy's teaching was highly selective, and visitation after visitation revealed that the progress of the Reformation was slower than the clergy and the state had expected. In the eyes of the religious and the secular authorities, the people's moral standard remained low and their knowledge about Christian faith was hardly sufficient.[55] For frustrated preachers, wonder was a convenient tool to catch their audience's attention and to press them to amend their lives, threatening them that God's wrath could befall them soon. A number of *Busspredigten* (sermons calling for repentance) mention *Wunderzeichen* to convince their hearers/readers that they are facing God's wrath, and virtually all books about celestial wonders, written by the clergy, are filled with lamentations over the sins and moral failures prevailing in Germany and intense calls for repentance.

Many clerical authors did not argue about whether the celestial phenomena they saw were signs of the end or those of temporary punishments, because they could mean both. In his pamphlet about *Fewerzeichen*, Hieronymus Opitz declares that 'without doubt' this is the sign of the imminent coming of the Last Judgement. However, he immediately adds that in the Scriptures, fiery signs often indicate God's wrath, which brings his punishment with a sword, and Opitz warns readers that Germany could be ravaged by the Muscovites if people do not repent.[56] There is no contradiction between these two interpretations because the Scripture states that prior to the Second Coming there will be various tribulations on the earth, including war, plague and natural disasters. As we saw before, Conrad Schomer regarded *Wunderzeichen* as the harbinger of both the Last Judgement and the disasters 'which Germany undergoes before the world's end'.[57] Since the Last Judgement is the time of final punishment of sinners, it is understandable that God would pour his anger on sinners prior to Christ's return.

Preachers often compiled a list of sins prevalent in society to urge the audience to consider the seriousness of these evil deeds in the eyes of God, but there were two special sins regularly condemned in *Wunderzeichen* discourses: people's ingratitude in response to God's grace and their false sense of security. Actually these two sins could be regarded as the sources of all evils. Lutheran pastors repeatedly thundered that German people were neglecting the gift of the Gospel, which God had mercifully bestowed upon them through Luther's Reformation. People did not take God's Word seriously and instead used the Gospel as an excuse to live in sin. Being filled with *Sicherheit*, they did not bother to improve their moral conduct. In the words of Nicholaus Erben, these stubborn people are dancing with the tune '*Pax* and *securitas*' until they end up in dancing the devil's dance in the abyss.[58] Surely God would inflict terrible punishment on such ungrateful, immoral Germans.

We have seen that Musculus, inspired by strange celestial phenomena in the sky over Frankfurt an der Oder, wrote a book on Luther's prophecy. In this book *Prophecy of Doctor Martin Luther* (*Weissagung D. Martini Lutheri*; 1568), Musculus reminds readers that in this last age, God had revealed his Word clearly to Germans, 'like the time of the Apostles'. No other land had enjoyed such privileges for the past fifteen hundred years.[59] However, he continues, Germans were abusing God's goodness and grace; despite repeated warnings, no moral improvement can be found in the people. Musculus asserts that there is no graver sin than contempt for God's Word, and such ingratitude will be punished with the harshest punishments: foreign dominance by the Turks, the Muscovites and the Tatars.[60] Here it should be recalled that the white beam of light coming from a mysterious rainbow was heading towards the land of the Muscovites. The implication of this phenomenon was clear: since almost all of Christ's (and also Luther's) prophecies regarding the End Times have come true, it is not strange that an invasion of Gog and Magog will befall Germany soon.[61] Now sincere repentance is the only way for salvation. Musculus repeats exactly the same discussion in his *Prayer to Escape from Deserving Imminent Punishment over Germany* (*Gebet Vmb abwendung woluerdienter zunahender Straff vber Deutschland*; 1569), invoking that many *Wunderzeichen* have been seen in the sky and on the earth within a short period, and 'we also know that in a land or city where such a sign appears, surely and always great misfortune, punishment and God's wrath follow'.[62]

Another example is Paul Eber's booklet on a strange celestial phenomenon that appeared in the sky over Wittenberg in March 1562. At 8 o'clock in the evening a white half-circle appeared in the northern sky, and large beams of light like 'long spears' flew all over the skies. Eber published an account of this phenomenon (with four crude illustrations which might have been drawn by his own hand), adding a lengthy admonition that readers should learn from this God-given sign. In this account he reminds readers how merciful God is, as his Word is now taught in churches, schools and books.[63] Indeed, God is so merciful that when he found that people did not listen to the preacher's words, God sent 'other co-preachers' ('andere Beyprediger') to give warnings to people so that they would repent – that is, *Wunderzeichen* in the sky and on the earth, such as eclipses, unusual movements of the stars, fiery signs and celestial apparitions.[64] 'Never in history, was there a time that so many ... dreadful ... signs appeared within short years'; in addition, 'the more [days pass], signs and wonders appear more dreadfully and frequently'. However, despite these supernatural warnings, the world becomes 'wilder, more malicious, and more rebellious, and more stubborn'. Eber thunders that people are tired of God's Word, 'living in *sicherheit*, abusing Christian Freedom, the Gospel, and faith to cover up and excuse all kinds of disgrace and evil habits'.[65] But such a sinful state of the world cannot last long; soon the dreadful end will come.[66]

In his book about the 1577 comet *Theological Reminder of the New Comet* (*Theologische erinnerung von dem newen Cometen*; 1578), Andreas Celichius presents a vivid image to impress people of the grave result of their sins. The Mecklenberg Superintendent first points out that natural philosophers are teaching that comets originate from grisly, sticky vapours and mists that come up from the earth, catch on fire in the sky and burn like sulphur. However, the true origin of comets is human sins: thousands of people's sins and disgraces go up to heaven like thick smoke filled with a disgusting stench. This mass is ignited by the fire of God's wrath and turns into a comet. This quasi-scientific theory of 'sinful vapour' as an origin of comets also appears in Heerbrand's sermon on the 1577 comet. Interestingly, Bapst, Johann Cuno and Petrus Victorius employed this theory to explain the origin of *Fewerzeichen* too. Cuno, who compiled a lengthy book on *Fewerzeichen* in 1578, quoted a Scripture passage which might have inspired this view. In Genesis 18, God tells Abraham, 'The outcry of Sodom and Gomorrah is indeed great, and their sin is exceeding grave. I will go down now, and see if they have done entirely according to its outcry, which has come to me'. Cuno adds the words 'the stink coming from Sodom' to 'the outcry' which reached God.[67] The strong image of the 'stink of sin rising up to heaven' symbolizes the preachers' view on the moral failures of the people.[68]

There are more examples of the Lutheran clergy's *Busspredigten* that quote celestial phenomena, but the main content of their messages is the same.[69] Thus *Wunderzeichen* gave clergy an opportunity to vent their anger and frustration with the slow progress of the 'Reformation of life' and the people's lukewarm attitude towards God's Word. One passage from Ambrosius Taurer's book, based on Christ's lamentation over stubborn Jerusalem (Matthew 23:37), summarizes the Lutheran clergy's frustration and sense of powerlessness: 'O Germany, Germany, [thus] speeches Christ through his servants [pastors like Taurer]: How often I wanted to gather your children together, the way a hen gathers her chicks under her wings, and you were unwilling. Behold, your house is being left to you desolate'. In the following lines, Taurer invokes the danger of invasions by foreign armies such as the Turks and the Muscovites, and reminds people that God has been threatening them through wondrous signs ('fiery rod, namely fiery light beams') for thirty years.[70]

However, there is also a theological reason why Lutheran preachers filled their messages with harsh condemnation of sins and threats of divine punishment. *Wunderzeichen* discourses had a function: to terrify people's consciences and lead them to the Gospel, based on the Lutheran principles of the Law and the Gospel. Luther, Melanchthon and their followers unanimously agreed that in order to understand the true meaning of the Gospel, one must confront God's Law (*Gesetz*) and recognize that one is utterly powerless to fulfil what God has commanded. This means that one cannot escape from God's wrath and eternal damnation by one's own power, and it naturally makes people fall into spiritual despair. However, by undergoing this painful process, one is led to the Gospel and finds the forgiveness of sins in Christ's cross. In Melanchthon's words, the 'beginning of a Christian life is to be frightened deep in our heart because of the wrath of God and our own sin'. This fear and despair are the new start: '[w]hen the frightened heart comes to understand that all our merits are not enough to atone for our sins, it turns instead to Christ and grows certain that he has lifted our sin from us'.[71] As Johann Wigand stresses, the Law and the Gospel cannot be useful if only one of them is taught. Those who want salvation must first be humbled through condemnation by the Law and confession of their sin. Only those who underwent this process can be truly consoled by the Gospel.[72] It is no wonder that in so many sermons, Lutheran clergy bombarded the audience with intimidating Scripture verses and historical examples of disasters that warned of God's wrath and punishment. They intended to terrify the hearers so that they would truly comprehend the reality of their fallen nature and turn to God for help.

It is clear that *Wunderzeichen* were a convenient tool to remind people of God's strict Law and wrath. In October 1570 a human hand, rods and other strange objects allegedly appeared around the moon in the sky over Hesse. An Erfurt broadsheet which reports this phenomenon starts with the statement that God calls people to repentance through 'the teaching of the Law ['der lehr des Gesetzes'], dreadful signs in the sky and fatherly punishments ... which people may bear'.[73] Here *Wunderzeichen* are clearly included in the same category as the preaching of the Law. There is at least one sixteenth-century religious discourse written by a clergy which explicitly uses the word *Gesetz* to explain the meaning of celestial signs. According to a devotional tract on the 1580 comet written by the Erfurt preacher Adelarius Praetorius, everything on the earth and in the sky can be categorized as *Zornzeichen* (signs of wrath) or *Gnadenzeichen* (signs of grace). For the unrepentant, *Zornzeichen* are a 'true, eyes-filling visible sermon of the Law' ('eine warhafftige Augen füllende sichtbare Gesetzpredigt') that forewarns of God's revenge. As the hearts and bodies of unrepentant unbelievers burn with evil lust, the heart of God burns with the fire of his wrath. God reminds people of the necessity of repentance not only through the preached Word on the earth but also through 'fire poles and fiery bars' ('Feurstangen vnd Fewerlattern am Himel', which means comets) in the sky as 'fiery preacher'. The conclusion of the sermons is 'Repent and believe the Gospel. God will no longer keep out the Last Judgement and the day of revenge'.[74] Praetorius's words indicate that the comet was interpreted as a *Gesetzpredigt* sent by God, which presses unrepentant sinners to open their hearts to the Gospel.

Few discourses use the word *Gesetz* directly, as Praetorius did, but in many cases the clerical authors' intention to treat *Wunderzeichen* as the Law is clear. In his book on *Fewerzeichen*, Babpst states that we should keep the celestial signs in our minds as 'visible, earnest sermon for condemnation and repentance, which God the Lord ... preached from the sky'. Through this, God intended 'to soften the godless's petrified and Adam-like hearts, which usually ignores all warnings and examples, and bring them to confession of their sin'.[75] This is the very function that *Gesetzpredigt* is supposed to fulfil.

The principle of 'the Law and the Gospel' reminds us of an important point – that striking people's conscience was by no means the final goal of the clerical authors of *Wunderzeichen* texts. The ultimate goal was to lead readers to the Gospel so that they could find peace and consolation in Christ. Preachers often encouraged people to consider God's love and willingness to forgive so that they would not fall into despair. In his sermon, Wolfgang Peristerus repeatedly emphasizes the 'great patience, grace and mercy' of God, who gives ample warnings to people through his servants and all kinds of 'special, remarkable and dreadful wonder and signs' so that they will not end up in damnation.[76] The very fact that God sent celestial signs signifies his love for humankind, and this point is stressed in several *Wunderzeichen* discourses, including Eber's tract, which we have already seen. Knowing that God's intention is not to damn sinners, Peristerus urged hearers to 'pray to God ... to turn away His righteous wrath and deserving punishments', which he warned of through *Wunderzeichen*, 'or, yes, punish us not for our damnation but for our improvement and our soul's salvation'.[77]

In his booklet on the mysterious white beams of lights appearing in the sky in the spring of 1562, Pfeffinger first presses readers with biblical warnings on sin and states that God calls the godless through the 'preaching of God's Law, all dreadful sayings of the holy Scripture and history, also all dreadful sign in heaven, on the earth, and water' to turn to God and amend their lives.[78] After threatening readers that God will punish them severely if they do not repent, he assures them that Christ's precious blood is a thousand times greater than the sins of this world. 'Therefore our dead Lord and Saviour Jesus Christ calls all sinners and female sinners so friendly, "Come to Me all who are weary and heavy-laden, and I will give you the rest"';[79] do not despair like Cain and Judas and abandon faith in Christ's forgiveness. At the conclusion of the tract, he states that one must distinguish between the Law and the Gospel carefully, and not fall into arrogance (by focusing only on the Gospel) or hopeless despair (by focusing only on the Law.)[80] Pfeffinger's tract amply shows how the mechanism of a *Wunderzeichen* sermon should work.

This assurance of salvation and consolation for those who repent can be extended to the clergy's discourses on the Last Judgement. Pastors usually preached about the coming of the Last Judgement in threatening tones, but they often reminded people of the truth that if they truly turn to God, they have nothing to fear; rather they should welcome joyfully the day of their final salvation from sin and liberation from the tribulations on this earth. In his sermon on the second Sunday in Advent, Martin Luther comments on Christ's comparison of the coming of the Judgement with the approach of summer (Luke 21:29–31), stating that '[b]y this he clearly teaches that we are look forward to the last day with as much joy and delight as all creation shows in spring and summer'. Indeed, Christ also says that 'when these things [apocalyptic signs] begin to take place, straighten up and lift up your heads, because your redemption is drawing near' (Luke 21:28). If one truly desires to be liberated from sin and to become holy, Luther assures, one does not need to fear the return of Christ because '[i]t come to set free from sin all who desire it, and you belong to that number'.[81] Pfeffinger, who regards the recent wondrous incidents as signs of the End Times, also stresses the 'summer' image of Judgement Day and encourages readers to consider that 'God's Son is coming to this world to do good to me, not to damn me, but to make me holy'. Whether they are dreadful or not dreadful, all signs show 'nothing but a shoot and a sign of my Kingdom of Peace, [and] summer, which is my perfect salvation'.[82] Cuno also states that for believing Christians, *Fewerzeichen* is a 'sign of joy' because one's sadness and tribulations will come to an end soon. Only for the godless is a fiery celestial wonder a sign of wrath.[83]

The image of the fiery fall of the world, predicted in 2 Peter 3 and often referred to in *Fewerzeichen* discourses, is surely a horrific vision. However, some authors emphasize God's mercy, which will protect his children even in the midst of such apocalyptic fire. Sigismund Suevus encourages readers to repent by stressing a maternal image of God, who protects his children from all evils. God holds repentant Christians in his arms, talks to them cheerfully, and removes the tears from their eyes, as a pious mother does to her small child in distress. As a mother gives her child 'butter bread, ginger cake, apple, peach, sugar and other pretty food', God nurtures believers both physically and spiritually with his Word and Sacraments. Likewise, God will protect his children 'on the day of wrath, in the dreadful storm', just as a mother protects her child from wind, rain and snow by covering the child with herself.[84] Such an affectionate maternal image of God shows that the spirituality of *Wunderzeichen* discourses did not only consist of gloomy threats of divine punishment and dry phrases of 'repent or disaster'. Consolation and encouragement based on God's love and mercy were also part of the clergy's message.

However, the fact that the element of *Gesetzpredigt* is predominant in most of these *Wunderzeichen* texts is undeniable. Unlike Pfeffinger, many authors focused only on the Law aspect of celestial signs, filling pages with words of lamentation, condemnation and intimidation, while they mentioned God's gentleness and mercy in just a few lines. Probably they were afraid that extensive references to divine mercy would soften the tone of their messages too much and lessen the shocking effects on people's hearts. They did not want to give people any more room for *Sicherheit*.

Thus *Wunderzeichen* discourses functioned as an auxiliary of *Gesetzpredigten* which challenged people's consciences in order to lead them to Christ's forgiveness. Celestial wonders served as an educational tool for people to carefully observe and meditate on, similar to their pastor's sermon. Suevus instructs readers about what they should do when a wondrous sign appears: a pious father should wake up his children, let them see the sign, admonish them seriously, and pray diligently. Likewise, as caretakers of God's mystery, pastors and preachers should always be awake and wake up their parishioners with the trumpet of God's Word. They should admonish people not to see God's Word and *Wunderzeichen* with 'dull eyes' ('schelen augen') but to pay attention to them as true warnings and serious *Busspredigten*.[85] Suevus's image of a father (both in household and church) showing his children celestial phenomena with admonitions gives us a glimpse of how celestial wonders were used to nurture piety in the sixteenth century.

According to Johannes Kepler's recollection, when the comet of 1577 appeared, his mother took him to the hill outside the city of Leonberg and showed him the sight.[86] Probably his mother admonished little Johannes to behave so that he could please God. Whether or not he remembered her words is unclear, but the experience became the starting point of Kepler's lifelong pursuit of God-given 'celestial harmony'. Thus his mother's educational effort at least worked to turn her child's attention to celestial matters, though it may have been a different type of piety from the one she had hoped.

Reassertion of the Clergy's Authority

When they saw dramatic *Wunderzeichen* in the sky, many people turned to the pastors, usually the most learned members of their community, for explanations. A printed broadsheet illustration about the fiery sky which appeared over Eckelsheim (December 1560) depicts this very scene: some villagers looked at the red-coloured sky in dismay, while a sexton rang the church bell. In front of the church, a bearded man talks to another man with a long robe and a black hat, and the latter is answering him, pointing to the skies with his right arm; judging from his attire, the robed man must be the pastor of the village.[87] Some clergy also recorded their lively exchanges with parishioners immediately after they saw strange phenomenon: according to Opitz, after they saw *Fewerzeichen*, 'many good-hearted people asked him' to write about this celestial sign and print it so that 'Gottes Wunderwerck / vnd Warnungs zeichenn' will not be forgotten.[88] People may have expected their pastors to preach about *Wunderzeichen* on the Sunday following the strange incident; when Wolfgang Peristerus published his sermon on the Sunday of *Reminiscere* in 1581, he attached a brief exposition on the *Wunderzeichen* witnessed the previous Saturday. Peristerus opens his description of the phenomena with the words 'Gesteriges tages (Wie jhr wisset)', which means 'yesterday, as you know'.[89] This suggests that Peristerus composed this exposition the day after he saw the strange signs and delivered it at his church during his Sunday sermon.

Actually, the clergy had ample reason to welcome opportunities to preach about *Wunderzeichen* because it gave them a chance to reassert their authority. In order to understand this point, we must consider the serious problems the Lutheran clergy had to confront in the latter half of the sixteenth century. Most of them had various troubles with their parishioners over their authority and working conditions. As Robert Scribner and other historians point out, the clergy lost part of their sacred authority through the Reformation: the power of transubstantiation, which can transform ordinary materials into sacred objects. Instead, the role as 'servants of the Word', the authority to interpret God's Word and to preach, became their new identity.[90] However, many pastors felt that people did not respect God's Word (and consequently, the servants of God) sufficiently. As we have seen, 'scorn of the Word' was a constant lamentation of the clergy, whose identity was founded on the authority of the Word.

What intensified the people's antipathy towards the clergy was the fact that the pastor depended on his parishioners financially. Parishioners were expected not only to provide for the living expenses of the pastor and schoolteacher but also to manage the buildings of church, school and parsonage, and to help in the pastor's ministry. Such expenses and labours were burdensome for many communities, and visitation records contain many conflicts between pastors and parishioners over these financial burdens. People often withheld these expenses to harass the clergy they disliked, and many pastors openly lamented their parishioners' unsupportive attitude.[91] Susan C. Karant-Nunn's study on the visitations in the rural regions of Electorate Saxony shows that despite the efforts of the Elector's officials and the clergy, people were reluctant to give up the traditional communal customs such as holding spinning bees and seeking advice of 'wise women', and they often showed their opposition openly.[92] Such attitudes of parishioners certainly strengthened the clergy's frustration with people's ingratitude towards the Word and its servants.

Considering this, it is not difficult to imagine that clerics welcomed *Wunderzeichen*. Shocked by frightful sights in the sky, people became more willing to listen to the preachers' messages. The preachers claimed that *Wunderzeichen* appeared as an ultimate preacher because people had not taken heed of the clergy's messages. For Simon Pauli, the 1577 comet was 'a fiery burning prophet and preacher' who was sent by God since people had not listened to the Word that had been preached by his servants for decades.[93] Now nature had started to show the consequences of the people's disregard for God's Word; it signified a final warning, and surely God's punishment would follow if people ignored this message. Georg Caesius includes a similar passage in his comet chronicle. According to the pastor of Leutershausen, the comet in 1577 had the shape of a broom or a large rod, and it showed God's will to punish people and to urge them to amend their lives. He immediately adds: 'Because of the *sicherheit* and neglect of God's word and the church staff are too great, God himself preaches through such sign from heaven'.[94] Thus *Wunderzeichen* gave the clergy an opportunity to reassert the authority of the Word and the servants of the Word.

The opening passage of Adelarius Praetorius's pamphlet on the 1577 comet, probably based on his own sermon, neatly summarizes this:

> History and also experience tell that this: unusual sign in the high heaven and in the sky indicates that God gives no forgiveness but great punishment for people's scorn of the Word and unrepentant attitude. It forewarns of ... plague, change in regime, war, inflation and devastation of lands and people. In all ages God has used not only the Word and preaching but also signs and wonder when He punishes the world for its sins, [and He] urges [people] to repentance and improvement of their lives.[95]

Then the Erfurt preacher invokes the well-known example of Jerusalem and the signs given to the city: As Jews scorned Christ's and the Apostles' warnings, God then preached to them through *Wunderzeichen* such as earthquakes, storms, eclipses, a comet and the apparition of a sword. But the people still did not take heed, and consequently the Jewish kingdom was destroyed.[96] Here the parallel between the Jews' contempt for the warnings of Christ and the Apostles, and people's contempt for the preachers' messages is obvious. Now a comet has appeared; what will follow next, if we consider the example of Jerusalem? The answer is clear.

After he saw the 1577 comet, Nicholas Selnecker compiled a prayer booklet which contained two prayers of confession to seek God's forgiveness. In these prayers, Selnecker describes the sins people should confess to God and repent of, and he does not forget to describe the people's sins against the clergy. 'To Thy faithful servants and workers ... we did not listen carefully but ... scorned and ridiculed them, their admonition, warning and condemnation'.[97] Failure to provide money and provisions to pastors, neglect of a Christian education for children and indifference to Christian discipline follow.[98] Obviously Selnecker believed that his parishioners had a lot of things to ask forgiveness for due to their ill treatment of God's Word and the clergy.

Indeed, celestial wonders provided an opportunity to present complaints about people's mistreatment of pastors. It is interesting to note that in his pamphlet about the 1562 *Wunderzeichen*, Eber complains about exactly the same kinds of troubles described above: he laments that people are not only ungrateful but also unkind and 'sometimes very very tough' ('bisweilen viel viel zu hart') towards teachers and pastors. They rob church provisions and incomes that pious people donated to maintain the preachers and worship services. Such actions force pastors to live in poverty, and their widows and orphans are left unsupported; the need for children's education is forgotten; buildings of churches and schools are repaired only temporarily at the time of visitation, and it is done with small expense.[99] As the General Superintendent of Saxony, Eber had to hear much about the difficult condition of the pastors and their ministry, especially in rural areas. Thus he used celestial wonders to remind people of their unfair treatments of the pastors and that God counted such acts as grave sins to be punished.

Conclusion

The above examination of *Wunderzeichen* in the clergy's messages shows that calls of 'repent or disaster' were not just empty stock phrases. They reflect the Lutheran clergy's deep resentment of the 'terrible ingratitude and disgraceful godless lives' ('grosses vndanckbarkeit vnd schendlichen Gottlosen lebens') of people, especially their scorn of God's Word.[100] What underlies this accusation is the sense that Germans were, in Musculus's words, 'the last Israelites' ('wir letzten Jsraeliter'), who had received exceptional divine grace and blessings through the preaching of the Gospel for decades. However, if the Germans are such a special people, surely their sins and ingratitude will be punished in the same way the Jews were punished. And in a number of *Wunderzeichen* works (including those by Praetorius), authors repeatedly reminded readers that various strange phenomena appeared in the sky and on the earth before the destruction of the Jewish kingdom. Thus for these ministers, stories of celestial wonders were convenient tools to express their frustrations about the slow pace of the Reformation of life and to attack people's *Sicherheit* and sinful lives. This regular denunciation of sins was in accordance with the Lutheran theological principle of the Law and the Gospel. By representing God's Law and his wrath on transgressors, celestial wonders could also be used as *Gesetzpredigt* to press people to recognize their desperate condition and seek help in the Gospel.

The clergy also used stories of wondrous phenomena to reassert their authority as God's servants, and in some cases *Wunderzeichen* were used as a chance to present complaints about their treatment in the community. Even if people scorned the Word and held preachers in contempt, nature could not ignore people's sins; nature showed the signs of divine wrath so that people could no longer ignore the Word. For preachers, nature and *Wunderzeichen* were strong allies in their ministry work, and Eber's referring to celestial wonders as *Beyprediger* reflects this sense well. Considering these points, it is understandable that many preachers showed great interest in wondrous phenomena and discussed them in their writings. They were educational materials to teach people both God's severe Law and his affectionate love. Because he does not want the damnation of sinners, God, as a 'faithful Eckhart', mercifully places warnings signs in the sky so that people will turn to God before it becomes too late.[101] For those who repent, celestial wonders were not the dreadful sign of God's wrath but an encouraging sign of divine grace, and they could wait for the return of Christ without fear.

We have seen that the backbone of the clergy's *Wunderzeichen* discourses is their eschatological conviction. Christ's prophecy concerning the signs of the End Times tightly connected celestial phenomena and the Last Judgement in the minds of the clergy. Without this biblical foundation, no matter how strong Melanchthon's academic influence was and how dramatic and shocking experiences of *Wunderzeichen* were, Lutheran *Wunderzeichen* discourses were not able to stand.

However, the above examples were by no means the only uses of *Wunderzeichen* in the clergy's writings. Strange celestial phenomena were not the only signs of the end that Christ foretold. He also warned that many false prophets would appear and mislead people, and the eschatological context of *Wunderzeichen* discourses gave some clergy a chance to use stories of wonder for polemical purposes. In the following chapters, examples of theologians who connected *Wunderzeichen*, the End Times and 'false prophets' to discredit their theological opponents will be examined.

3 'AN EAGLE HURTING HIMSELF': FLACIUS'S TRACT AGAINST THE INTERIM

In June 1546, only four months after Martin Luther died in Eisleben, a disaster befell the Lutherans in Germany. While hiding his true intentions, Emperor Charles V launched a military campaign against the German Protestant princes to solve by force the problem of Christendom's split. Duke Moritz of Albertine Saxony, a cousin of Elector Johann Friedrich of Saxony, took the emperor's side despite his Protestant faith, and his invasion of Ernestine Saxony aggravated the situation in the Lutheran camp. On 24 April 1547, surprised by the imperial army at Mühlberg, Johann Friedrich's force was completely crushed and the Elector himself became a prisoner. The defeat at Mühlberg was an apparent fatal blow for the Protestant cause. As the Landgrave Philip of Hesse surrendered and became a prisoner, the resistance of the Schmalkaldic League immediately collapsed, and Wittenberg also capitulated to the imperial army in May 1547. Next spring, the victorious emperor tried to impose his provisional religious settlement, commonly called the Augsburg Interim, on Protestant territories in order to restore Catholic practices. Soon the Lutheran camp split into two groups over the question of whether they should keep fighting against the emperor's religious policy or make concessions on some theological points and come to peace with him. Those who chose to compromise formulated a new religious settlement nicknamed the 'Leipzig Interim' to appease the emperor, but the opposing group strenuously rejected it as a hideous compromise with the Antichrist. This tragic division became the starting point for the long theological controversies which plagued the Lutheran Church for over the next thirty years.

This chapter examines a unique *Wunderzeichen* tract published by the most prominent opponent of the Interim, Matthias Flacius Illyricus, and analyses how stories of wonder were used by the Lutheran theologians during these critical years of the Interim Crisis (1548–52). In order to understand the nature of Flacius's tract, we will take a brief look at the history of the Interim Crisis and consider why it caused such a serious split within the Lutheran Church. The important point is that Flacius and his colleagues interpreted this current crisis in an eschatological light. They regarded the struggle over the Interim as an apocalyptic battle between Christ's Church and Satan and urged people not to trust pro-Catholic settlements, which were nothing but the Antichrist's trick to eradicate evangelical faith from the earth. Flacius's pamphlet on celestial apparitions was also founded on this apocalyptic conviction. An analysis of the symbolic meanings of the mysterious aerial apparitions described in the text, such as the double-headed eagle, the weeping maid and an image of Elector Johann Friedrich, reveals that the pamphlet, obviously intended for wide circulation, served as a propaganda weapon for Flacius's campaign against the Interim. It shows that the stories of the Papal Ass and the Monk-Calf were not the only examples of the propaganda use of wonder stories; news of celestial phenomena could also be used for polemical purposes. Melanchthon's interest in wonders is also re-examined in this chapter, and it reveals that he too sought political and theological messages in heavenly phenomena during the stormy years following the Schmalkaldic War. However, his interpretations of celestial phenomena show a marked difference from Flacius's, as his role in the Interim Crisis differed from that of his former colleague.

The Interim Crisis and Flacius

The doctrines presented in the Augsburg Interim were hardly acceptable to most Lutheran ministers. Although embellished with various flowery Lutheran stock phrases, the decree virtually denied the theology of 'justification by faith alone' and other key doctrines of the Protestant faith. While it declared that God justified human beings without their merit, the significance of good works is emphasized as 'necessary for the salvation of everyone justified', and without them one cannot keep God's grace with himself. 'Through cooperating with grace', those who are justified 'abound in good works, and the more they abound in them, greater growth in righteousness is added to them, so that those who are righteous become more righteous'.[1] The Interim also demanded restoration of virtually all Roman Catholic rituals, including prayers to saints and mass for the dead. Despite the quasi-Protestant statement that mass is celebrated in thanksgiving for God's mercy, the ritual was still defined as a sacrifice, and the doctrine of transubstantiation was affirmed. The only concessions to the Protestant side are a temporary recognition of clerical marriage and the communion in two kinds. It is no wonder that Philip Melanchthon, now the leader of the Lutheran Reformation after Luther's death, and other theologians of Wittenberg University strenuously refused to accept such a settlement.

Moritz, who received a large part of the Ernestine Saxony (including Wittenberg) from the emperor as a reward for his loyalty, had to admit that he could not introduce the Augsburg Interim without risk of a general rebellion. However, unable to ignore the emperor's will, Moritz and his councillors put heavy pressure on the Wittenberg theologians to formulate a compromise that could satisfy both the emperor and the Lutherans. Melanchthon was terrified by the news that hundreds of Lutheran pastors were expelled from the southern German cities occupied by the imperial army. Rejecting all compromises with the emperor would lead to a mass exodus of Protestant clergy, and this meant that the inhabitants of these territories would be deprived of God's Word. Wittenberg University, the citadel of Lutheran teaching, would surely be closed. Fearing that Luther's teachings would be extinguished and people would lose their spiritual shepherds, Melanchthon, Johann Pfeffinger, Georg Major and other Wittenberg theologians finally agreed to cooperate with Moritz. The proposal of settlement which they drafted, the so-called Leipzig Interim, was presented at the diet of Saxony in December 1548, though it was never officially published.

Compared to the Augsburg Interim, the doctrine of justification by faith in God's grace is much stressed in this document, but it deliberately avoids the phrase 'through faith alone'. In order to justify the restoration of Roman Catholic rituals, Melanchthon introduced the concept of adiaphora (which means 'indifferent thing' in Greek) – things that are neither commanded nor forbidden by God. Catholic ceremonies could be reintroduced because they were not explicitly forbidden in the Bible and do not affect the foundational doctrines of Lutheran faith. Melanchthon insisted that he had managed to retain the essence of evangelical faith, and that there was no teaching which deviated from Scripture and Luther's writings in this settlement. Some historians claim that Melanchton's use of the concept of adiaphora does not contradict Luther's church policy, which sometimes showed a remarkable degree of flexibility.[2] However, to modify the speed of the Reformation's progress in predominantly Catholic territories and to restore Roman Catholic practices in already reformed territories in order to satisfy the secular power were totally different matters. Many ardent Lutheran theologians were not impressed by Melanchthon's arguments about adiaphora. They severely denounced the Leipzig Interim as a coward's concession to the Antichrist which would eventually lead people to total apostasy from God's Word.

Since then, Melanchthon has been condemned by many church historians and theologians as an indecisive compromiser who endangered the purity of the Lutheran faith. Recent historians are less harsh towards Melanchthon's desire of peace and openness towards dialogues. In his recent article, Timothy J. Wengert claims that Melanchthon even showed his readiness to become a martyr at the beginning of the Interim Crisis.[3] Robert Kolb also emphasizes that Melanchthon was 'a man of considerable courage, a strong and skillful confessor of faith' and points out his refusal to compromise to a greater extent in spite of the pressures from the secular advisors of Moritz.[4] However, the traditional negative image of the timid Preceptor still remains influential, as Oliver K. Olson's book – which even suggests the possibility that Melanchthon was bribed by Moritz –

amply shows.[5] If Melanchthon failed to win the confidence of the historians of later ages, it is easy to imagine that how mercilessly he was denounced by other Lutheran theologians, living in the tense atmosphere after the military defeat. For them, the fact that Melanchthon, a long-time comrade of Luther and author of the *Augsburg Confession*, yielded to the power of the Antichrist was nothing but a nightmare. Hundreds of anti-Interim and anti-'Adiaphorist' pamphlets were published to urge people not to be deceived by the 'sophist philosophy' of the Wittenberg theologians. How can they claim that they are still preaching God's Word while they allow restoration of the Roman ceremonies that opposed God's Word?[6] True believers should choose suffering rather than temporal peace with their enemies, they insisted. The theologians who produced these tracts were called Gnesio (genuine)-Lutherans because they regarded themselves as the only true heirs of Luther's theology, in contrast with the Philippists, the followers of Melanchthon.[7]

A number of Lutheran theologians who lost their positions due to the emperor's policy headed to Magdeburg, a city that dared to fight against the Interims. Thanks to prolific authors among its refugee theologians, Magdeburg soon became a publication centre of the anti-Interim/anti-Adiaphorist campaign and won the nickname of 'Our Lord God's Chancery'. Charles imposed an imperial ban on the city in July 1547, but the clergy and magistrates of Magdeburg claimed that the city had a right to resist the emperor because he had intervened in religious affairs, a realm in which he did not have any real authority.[8] Though the new Elector Moritz besieged the city in 1550–1 to carry out the emperor's ban, the city endured the siege and forced Moritz to make peace under relatively generous terms.

The sheer number of Lutheran books produced in Magdeburg for propaganda purposes in these years is indeed striking. Although Magdeburg did not have a long history of printing when the Interim Crisis started, during the two early years of the crisis (1549–50) the city's publishers printed 206 titles, most of them propaganda works against the Interims. In total at least 360 works were published between 1548 and 1552. No other place in Germany ever published so many books within such a short period.[9] According to Thomas Kaufmann's estimation, almost 25 per cent of the entire book production of sixteenth-century Magdeburg was published in 1548–52.[10]

The Magdeburgers' publications seriously annoyed Moritz and hampered his effort to restore order in the newly acquired Saxon territories. Though he had to satisfy the emperor by implementing his designs, Moritz knew well that he was not a popular ruler and should not further alienate the populace by imposing pro-Catholic policy by force. Naturally he desired to avoid a costly military campaign against the zealous city, as he knew the difficulty in securing sufficient funds and soldiers in Saxony. People's sympathy for Magdeburg became even clearer as the siege started. Moritz's call for funds to support the campaign was rejected by the provincial diet, and the recruiter of his troops found that 'no one wanted to serve against Jerusalem and the Lord God's Chancery'.[11] Thus Magdeburg's propaganda campaign effectively undermined the political ambitions of Charles and Moritz and enabled the city to survive the storm.

The Gnesio-Lutherans regarded the conflict over the Interims as an eschatological and cosmic battle between the Church of Christ and the forces of Antichrist, supposed to take place before the world's end. The persecution of Lutheran pastors and the hideous 'apostasy' of Melanchthon and other Wittenberg theologians were regarded as fulfilment of the apocalyptic prophecy in the Book of Revelation. It was clear that Satan was waging a frontal attack on the Gospel and on true believers – surely the day of the Last Judgement had to be near![12] Now was the time where everyone had to choose a side and to take action. As one Magdeburg pamphlet printed in 1549 stated, '[a]ll Christians, on pain of the eternal curse and damnation, are required ... upon seeing their fellow Christian brothers in need, not to abandon and forsake us – indeed, not only this, but also to come to our aid'.[13]

This apocalyptic interpretation of the Interim Crisis is an important factor in the writings of Matthias Flacius Illyricus, the former professor of Wittenberg. Born in Croatian Illyria, he studied at Wittenberg and later became professor of Hebrew at the university. From the beginning of the Interim Crisis, Flacius strongly called for staunch resistance and pressed Melanchthon to stand steadfast against the emperor's religious policy. Disappointed by the peace-seeking attitude of Melanchthon, Flacius left Wittenberg in the spring of 1549 and moved to Magdeburg, where he became an energetic leader of the city's anti-Interim campaign and produced a massive amount of propaganda works.[14] Before his arrival at Magdeburg, Flacius published *A Prophecy and a Beautiful, Wonderful Consolation for All Lamenting Pious Christian Hearts in this Present Tribulation Time, Based on the 14th Chapter of the Revelation of John* (*Eine Weissagung vnd ein schöner Herrlicher trost für alle hochbetrübte frome Christliche hertzen zu diser jtzigen tru[b]seligen zeit Aus dem XIIII. Cap. Der offenbarung Johannis*; 1548), a tract in the form of a commentary on the fourteenth chapter of the Book of Revelation. The author's intention was to denounce those who accepted the Interims as false Christians who betrayed the true God and sided with Antichrist. Quoting the Revelation's passage of the Lamb being worshipped by 144,000 believers at the top of the mountain of Zion, Flacius reminds readers that while there are 144,000 true believers, there are many more people who reject the Gospel. Two kinds of churches exist in this world: the church of the pope, which does not stand with the Lamb and rather persecutes God's Word, and the church of the Gospel. The call of the first angel of the apocalypse to 'fear God and give Him glory' (14:7) tells Christians that they must not accept the 'hellish Interim' which removes the fear of God from their hearts. Christians must fear God only, not 'human being, emperor, king, and all authorities'.[15]

The second angel's declaration – 'Fallen is Babylon the great' – is a message of punishment for the false and apostate Christians who left the true teaching and accepted the 'blasphemous' Interim. Although they heard the Gospel for many years, they left God and God's Word and received 'the sign of the pope, which means his teaching and worship'. For this sin, the angel says, God's wrath befalls them and they are thrown into the fire.[16] The third angel's words – 'the smoke of their torment goes up forever and ever' – shows their eternal death and the right punishments for them. The next line, 'they have no rest day and night', signifies a strong contrast between true believers' peace with God and the despair of the unbelievers. Those who fell from truth and returned to the pope have 'no rest day and night' as they know that they did wrong to God. Driven by Satan, 'they commit one sin after the other, with persecution and murder of true Christians. Then they left the recognized truth and became liars, and finally turned into a murderous child ['Mordtkindt'] like their father [Satan]'. The book concludes with a prayer for God's protection, the quick return of Christ, the deliverance of believers from the devil's empire and punishment of the unbelievers.[17]

Flacius repeatedly used this allusion to Revelation 14 in his other anti-Interim/anti-Adiaphorist works, and he escalated his aggressive tone as the relationship between Gnesio-Lutherans and Philippists deteriorated. In his *Clear Proof that the Scripture is against the Interim and Adiaphora* (*Kleriche beweisung ... welche die schrifften widder das INTERIM vnd Mittelding*; 1550), Flacius accuses the adiaphorists ('or the masters of the noncommittal') of joining the whore of Babylon and her beast in spite of their knowledge about Antichrist, and now are persecuting Christ himself. They may claim that they have never persecuted Christ, but such a 'sophistic excuse' does not work, as Christ tells: 'Truly I say to you, to the extent that you did it to one of these least of brothers of Mine, you did it to Me' (Matthew 25:40). Certainly these persecutors will eventually be thrown into the fiery pool of eternal damnation. Flacius states that he trembles when he thinks about the German people's ingratitude towards the Gospel. In no country other than Germany is the Gospel known well; but at the same time, there are no people other than 'dear Germans' who neglect it so much. Undoubtedly, terrible punishment will befall this land after such 'dreadful imprudence and perversion', followed by eternal hellfire.[18] In conclusion, Flacius urges that while they still can, Christians should flee from the fire of God's wrath which burns *Interimisten* and *Adiaphoristen*.

These messages clearly show Flacius's conviction that the current crisis was the fulfilment of biblical prophecy of the End Times. For Flacius and other Magdeburg theologians, it was important to proclaim the imminent coming of the Last Judgement as it explained to people why pious believers had to suffer from such persecutions and also gave them courage to stand firm against the pressure of the Interim. At the same time, they believed that it was their duty to warn people not to endanger their souls and invite God's wrath by trusting the 'sophist' theology of Wittenberg professors.

As we saw in Chapter 2, Lutheran theologians showed great interest in Christ's prophecy of the end in Luke 21 and applied it to the *Wunderzeichen* of their age. Did Flacius and his comrades also seek proof of their convictions in extraordinary phenomena in nature? The answer is, they did.

Flacius's *Wunderzeichen* Pamphlet (1549)

While Flacius published two tracts about wonders during the Interim Crisis, neither of them was written by Flacius himself. He added introductions to these works written by others, but it seems that the initiative of publication came from him. One of these is *A Gruesome Sea-Wonder which Means the Pope, Found in Rome* (*Ein grausam Meerwunder den Bapst bedeutende zu Rom gefunden*), the famous account of the Papal Ass, a strange monster allegedly found dead on the riverbank of the Tiber in Rome in 1496. This pamphlet was originally written by Melanchthon in 1523 as blatant propaganda against the papacy, which interpreted the deformed body of the beast as physical proof of the monstrosity of the papacy and its corruptions. The fact that it was found dead in Rome was regarded as a prophetic sign that the papacy would soon come to an end. This text became a classic of Lutheran propaganda literature, and the fact that Luther reprinted it in 1535 and mentioned the story in his writings several times may attest to its worth as a propaganda weapon.[19] Flacius republished it from Christian Rödinger's press in Magdeburg in 1550, but in the preface he wrote for this tract, he does not forget to add some new biting comments about the original author, Melanchthon.[20]

The other *Wunderzeichen* work is *Sign in the Sky Recently Seen near Braunschweig, Written by the Superintendent of Braunschweig* (*Zeychen am himmel bey Braunschwig newlich gesehen durch den superatendentem zu Braunschwig geschriben*), published by the printing shop of Pancratz Kempff in 1549. The author of the text is Nicholas Medler, the Superintendent of Braunschweig and Melanchthon's friend. Probably the report had been published in Braunschweig or elsewhere before Flacius's edition appeared, but unfortunately it seems that the original version no longer exists. The tract consists of an illustrated title page, two illustrations, a five-and-a-half-page preface by Flacius and a five-and-a-half-page account of *Wunderzeichen* by Medler. On the title page, an elaborate illustration depicts a scene of four men in a travel wagon; one of them is pointing towards the sky and calling the others to look up.[21] The signs that appear in the sky are Christ on the cross between the two thieves, three moons, a lion, a two-headed eagle, a man with a sword and a maid begging for mercy (see Figure 3.1). Under the illustration there is a Scripture verse from Luke 21 that was frequently quoted in *Wunderzeichen* literature: 'When these things begin to take place, straighten up and lift up your heads, because your redemption is drawing near'.[22] The second page shows an illustration of Elector Johann Frederick with the date '1548'.

Zeychen am himmel bey Braun-
schwig newlich gesehen / durch den supera-
tendentem zu Braunschwig geschri
ben Mit einer vorrede.

Luce 21 wenn diss anfehet zu geschehn so sehet vff/
vnd hebt ewer heübter vff / darumb das sich ewer
erlösung nahet.

Figure 3.1: The illustrated title page of N. Medler and M. Flacius Illyricus, *Zeychen am himmel bey Braunschwig newlich gesehen* (Magdeburg, 1549). Herzog August Bibliothek Wolfenbüttel: H 92.4° Helmst. (14). Reproduced with permission.

The topics of Flacius's preface can be divided in three parts: the basic principles of Lutheran faith; the errors of the papists and their party; and the meanings of *Wunderzeichen* as signs of God's wrath and Judgement. Flacius opens his preface with the statement that throughout the history of the Church, in no other time than 'this last time' is God's Word manifested so clearly. He encourages readers to consider the difference between the Law and the Gospel, the 'key of the Holy Bible'. First, through God's Law, one must recognize his own sinfulness and God's wrath. Though this process terrifies and leaves one in sorrow, 'by faith alone' we find that only the Lamb of God can remove our sins. In this way one is liberated from Satan's power and becomes a child of God. However, since old damned Adam is still remaining, one commits sins daily out of weakness. Therefore, every day one must leave sin and live with faith in Christ.[23]

After this simple summary of Lutheran faith, Flacius explains that it has become clear that the papists have mingled philosophy, human laws and human reason with God's Word. The identity of the Antichrist has been revealed. People are taught the contents of the Scripture carefully, and nowadays even a ten-year-old child can explain them better than learned doctors in former times. Therefore, the author concludes, now is the 'golden time' when God pours his Spirit richly onto youth, and the knowledge of God fills the world like water in the sea. However, the world turned against this 'great, unspeakable Grace of our loving God'. How so? Here Flacius turns to two groups who are enemies of the Gospel: the first group consists of heathens like the Turks and the Jews; the second consists of the papists, who slander and persecute Christ and his Gospel and deliberately commit sins against the Holy Spirit. We know, Flacius confirms, that only the Lamb of God can remove our sins, and no other name is given to us for our salvation. For the papists this truth is only a heretical lie, and they insist on relying on mass, vigils and indulgences. They regard the communion in both kinds and clerical marriage as dreadful heresy (these practices are permitted in the Augsburg Interim, but obviously Flacius did not trust such promises; most likely he wanted to tell readers that in their hearts the papists do not recognize these Protestant practices, no matter what the Interims states). The papists also insist that seeing Christ as the sole mediator between God and human beings is a heretical idea and urge people to seek the help of the saints. Antichrist is also persecuting the Word of God and the Holy Spirit with his associates and 'the children of darkness'. Some of these associates had once 'tasted the Holy Spirit' but fell away to Antichrist. Feeling too ashamed to deny Christ outright, some devised clever means so that they could serve two masters.[24] Obviously the first group of Antichrist's *gesellen* (associates/company) signifies those who accepted the thinly veiled Roman Catholicism of the Augsburg Interim, while the latter represents the adiaphorists who tried to maintain their Evangelical faith by making theological compromises represented by the Leipzig Interim.

Flacius then asks his readers what God would think about such 'great, devilish ingratitude of the world', stating that God, as a stern judge, will certainly punish 'apostate *Mameluken*'.[25] God proclaims his will not only through preaching but also through 'many incidents in the skies, the water, and the earth'.[26] For example, Flacius mentions that mysterious springs of blood appeared in Silesia, Franken, Saxony and Meissen, he and connects these cases with the execution of the seven Magdeburgers who tried to kidnap a wealthy Jew in the enemy side:[27] 'The earth shows that the innocent blood of Abel cannot be concealed'.[28] Many other strange phenomena were seen in nature, and as Christ foretold, they are signs of the imminent coming of the Last Judgement.[29] Flacius closes his preface with the statement that these signs are God's call for repentance and at the same time a consolation for Christians in sorrow, as the following narrative of the Braunschweig case shows.

In this short but well-organized preface, Flacius covers all the important basic points that one should know during this time of confessional crisis: what the core of Lutheran faith is; why papist teachings should be avoided; why Christians are persecuted in this age; and what the strange signs signify. Though the main target of Flacius's condemnation is the papists, those who tried to reconcile with the papists by 'clever devices' are also condemned as the *gesellen* of Antichrist and the apostates, because both of these groups persecute the Gospel and true believers. Flacius's repeated emphasis on the persecution of the Protestants and its connections with *Wunderzeichen* in recent days gives readers a clue as to how to 'read' the news of strange apparitions in the following section of this tract.

The main narrative of the Braunschweig case, written by Medler, opens with a description of the characters of the witnesses. Three 'honourable and pious' merchants of Braunschweig had been staying at Antwerp, but they desired to return to Braunschweig by Pentecost to listen to an evangelical sermon on this holy day – unfortunately, evangelical sermons were still rare in Antwerp (indeed, Antwerp was a centre of Protestant persecution in the Netherlands). They continued their travel until midnight because they were afraid to miss the Pentecost sermon, and even during travel they discussed God's Word, thought about the meaning of the Pentecost, and sang 'Veni Sancte Spiritus' ('Come, Holy Spirit').[30] These passages establish the credibility of the witnesses by emphasizing the deep piety of the three merchants. Since night travel was dangerous and hardly common in the sixteenth century, Medler possibly felt the need to explain why these merchants were travelling on a lonely road at midnight if they were really 'honourable and pious'. In addition, these lines show that they were not drunk.

On the road between Celle and Braunschweig, they suddenly saw strange phenomena in the sky. First, the merchants saw a bright halo around the moon, followed by the three moons in the sky. One moon was shining brightly while the two others were rather dark.[31] Then a big, fiery lion appeared, followed by 'an eagle with two heads, one head is dark while the other is clearly visible'. The eagle is hacking his own breast 'like a pelican'. Then a big, wonderful picture of the imprisoned Elector Johann Frederick (the author adds the phrase 'gott helff im gnediglich' after the Elector's name) was seen. An image of Adam and Eve, with God placing his hand on them as a mother would, also appeared. After this strange sequence of apparitions, many animals appeared in the sky, including one camel that was observed clearly, followed by various visions of crowds of people, towns and castles. Christ on the cross between the two crucified thieves and a gathering of apostles also appeared. The apparition at the end is described as a big, fearful man who raises a long, mighty sword with both hands as if to strike, and a woman who weeps bitterly, prays, and raises her hands.[32]

Two of the merchants saw the celestial visions first and awoke the third merchant. When the third woke up, he saw a camel in the sky and was greatly terrified. There was also a boy ('knabe', a servant of the merchants) in the wagon, but it seems that they left him sleeping. The three merchants were so shocked that they believed that the Day of Judgement had finally arrived, fell on their knees and prayed earnestly to God that he would not punish Braunschweig severely. They also determined that once they reached Braunschweig, they would warn people and admonish them to repent. Soon they arrived at the gate of the city, but it was still early and the gatekeeper refused to open the gate. Out of excitement and fear that God's Judgement would start immediately, the merchants could not wait until dawn, banged on the doors of nearby houses outside the wall, awoke the residents and urged them to pray and repent. The news spread through the whole of Braunschweig immediately, and Medler, as the Superintendent of the city, launched an investigation. As he talked to two of the witnesses of this celestial vision, he confirmed that they were mentally sound and he could not find even the slightest sign of melancholia in them, which means that the vision they saw was not the product of an unbalanced condition of the four humors.[33]

Medler also received news that a man of Leipzig named Andreas witnessed a dreadful thing ('schreckliche ding') in the sky on the road between Braunschweig and Lübeck during the same night, and the contents of his apparitions are even stranger: 'On the cross there hung [a man] not like Christ's image because he had a gray beard', and his head was then beheaded by a sword. Besides this enigmatic apparition, two angels and 'many other things' were observed. In his conclusion, Meddler states that he can only write one thing: that this is a merciful warning from God of his Last Judgement and a call to turn to Christ.[34] A woodcut image of Christ holding a globe is printed on the last page.

We do not have information about the circumstances of publication of this tract, but it is hardly believable that Flacius printed it simply to entertain readers with a story of wonder. Virtually all of Flacius's writings published during the Interim Crisis are anti-Interim/anti-Adiaphorists propaganda, and he never involved himself in the production of *Wunderzeichen* literature before or after the publication of this tract. Moreover, the fact that Kempff published at least four versions of this text shows that Flacius and Kempff intended it for a wide circulation outside of Magdeburg.[35] Unlike the prominent Magdeburg printers Michael Lotter and Christian Röddinger, Kempff was running only a small-scale print shop, and he had no reason to spend money to print a non-propaganda piece while the whole city was expecting an imminent military attack. Moreover, Kempff was well known as a printer of elaborate anti-papist broadsheets, including the first known broadside song against the Interim.[36] In addition, considering the political situation of Saxony and Magdeburg in 1549, news of the apparition of the portrait of Johann Friedrich, still imprisoned because of his steadfast refusal to accept the Interim, cannot claim to be apolitical. Considering these facts, we can safely assume that this tract was a part of a propaganda campaign by Flacius and contains both political and theological interpretations.

In the Braunschweig case, a number of visions appear in the sky one after the other like a magic lantern. Is it possible to reconstruct how contemporary people interpreted the apparitions of the Braunschweig *Wunderzeichen*? There are at least three objects of which we can analyse the meanings based on contemporary sources: the eagle, the weeping maid and the image of former Elector Johann Friedrich.

The Eagle and the Maid

How is it possible to decipher the meaning of the eagle? One type of material that gives us a clue is sixteenth-century prophetic literature, filled with animal symbolism. As Sharon L. Jansen's study of English prophetic literature under Henry VIII's reign shows, many 'prophecies' that describe animals making alliances with or fighting each other were circulating among the populace. Common people seriously searched for political meanings in these enigmatic passages by applying animal symbols to contemporary political leaders.[37] The prophecy of Johann Carion, a schoolmate of Melanchthon at Tübingen, was one of those works. It appeared in the 1520s and was reprinted several times during the years following the Schmalkadic War, obviously because the latter half of the prophecy was believed to describe the political conditions of Germany after the war. Carion's prophecy narrates various interactions between eagles, lions, dogs and other animals. An exposition of Carion's book published in Strasbourg in 1548 explains that the eagles in his prophecy should be interpreted as the Habsburg emperor and his sons, in this case Maximilian I and his grandsons, Charles and Ferdinand.[38] This identification of the eagle as the Habsburg prince was common because the Habsburg herald was a double-headed eagle, and it was displayed in every place in Germany as the symbol of the imperial power.

Thus, most likely, the double-headed eagle of the Braunschweig *Wunderzeichen* symbolizes Emperor Charles V. If so, what does the apparition say about him? According to the text, one of the heads of the eagle is half-dark while the other side is clear, and this head was hacking its own breast 'like a pelican'. In Flacius's time, the pelican was commonly depicted as the symbol of Christ and charity, as the bird was believed to feed her children with her own blood pierced from her breast. At first sight, correlating the eagle and the emperor who persecutes Protestants with the symbol of charity seems odd.

However, there is one interesting fact. On the right side of the passage that reads 'gleich eim pellicann in die brust gehacket' ('like a pelican hacking the breast') is printed a marginal note: 'verderbet sich selbs'.[39] Marginal notes are common in sixteenth-century books, and usually they show a summary of paragraphs or the chapter numbers of biblical quotations. Strangely, these three words are the *only* marginal note in this tract, and probably it was deliberately added by Flacius to clarify the meaning of this passage. *Verderben* means 'hurting' or 'corrupting', but it can also mean 'damning'. Why is the eagle 'hurting/damning himself'? In the preface of the tract, Flacius repeatedly emphasizes the papists' sin of persecuting evangelical Christians. If the eagle represents the emperor, most likely it means that Charles's religious policy aiming to restore Roman Catholicism is hurting both himself and true faith, if not damning his own soul. The reason the marginal note has been added to the line 'like a pelican he is hacking his breast' is probably that by this note, Flacius wanted to make sure that this eagle was not acting out of love like a pelican but 'hurting' and 'damning' himself. At the same time, the marginal note suggests that the eagle's strange behaviour is important in this text because it is the only section where a marginal note is attached.

Before the eagle appeared, a lion was seen by the Braunschweig merchants. Several lions are mentioned in Carion's prophecy and interpreted as various powerful princes, including Landgrave Philip of Hesse. Since the herald of Braunschweig was a lion, it could also signify the home town of these merchants. Indeed, Braunschweig refused to accept the Augsburg Interim and fell into serious trouble with the territorial prince, Heinrich the Younger of Braunschweig-Wolfenbüttel, a staunch Roman Catholic duke who had been restored as a result of the Protestant defeat. A camel does not appear in Carion's prophecy, but it is commonly regarded as the symbol of the Turks in *Wunderzeichen* literature (visions of a camel were often seen in the sky, and several wonder books and broadsheets recorded these).[40] Judging from Melanchthon's comment on the Braunschweig case (which we shall examine in a later section), a camel foretold God's punishment of sinful Germany, probably by an invasion of the Turks. As we saw in Chapter 2, many Lutheran clergy, including Luther himself, seriously feared that God would allow the Turks to invade Germany as his punishment for people's ingratitude towards God's Word. The apparition of a camel might be a reflection of such a concern.

The last apparition in the Braunschweig case is a fierce-looking man with a long sword and a weeping maid. From the text, it is not clear what the man and the maid are doing, but the illustration printed on the title page of the tract shows clearly that the man is about to strike her with a sword and the maid is begging for mercy. Since the Church was often depicted as a virgin or bride of Christ, this scene could easily be interpreted as an allusion to the persecution of the Protestant Church. There is one contemporary voice to suggest that sixteenth-century readers associated this image with the Church's suffering during the Interim Crisis. In his wonder book *Wunderwerck vnd Wunderzeichen Buch* (1557), Caspar Goltwurm briefly describes the Braunschweig case and adds a comment on the image of the weeping maid, 'a very sad image':

> In these years, the holy Christian Church, the spouse or the bride of Christ, was terribly saddened and terrified so much. Therefore she [the maid in vision] lamented and wept out of sorrow. During this time Christ was also condemned falsely and put on the cross by our high priest the pope, bishops, and others.[41]

Here the image of the weeping maid is directly associated with the persecution of Lutherans brought about by the Interim Crisis. Goltwurm himself shared suffering similar to that of Christ's bride: he was expelled from his position as the Superintendent of Nassau-Weilburg in 1550 because of his active role in opposing the Interim.[42] Considering his book appeared only eight years after the publication of Flacius's tract, we can assume that many readers of Flacius's tract interpreted the image of the maid as Goldwurm did.

Perhaps some contemporary people identified the image of the poor maid with the city of Magdeburg. The herald of the city is a young maid ('Magd') with a flower wreath standing on the firm city wall, and in the anti-Interim literature Magdeburg is frequently represented as a chaste maid who refused to sell her own body to the pope and the Interims.[43] By the time Flacius's pamphlet was published, the city had already been under the imperial ban for two years and could be attacked at any time. It is not strange that people saw a parallel between Magdeburg and the image of the maid about to be struck. In addition, it is possible that the weeping maid symbolizes both the persecuted Church and the pious city at the same time. After the siege, Erasmus Alberus, one of the theologians who worked for Magdeburg's propaganda campaign, wrote: 'Magdeburg, the holy and worthy city, has a virgin for its seal; this signifies holy Christendom, which God strengthens and succours in its trouble'.[44] Here Goltwurm's interpretation of the apparition and Alberus's imagination overlap.

The meanings of other apparitions – the town, the castle, Adam and Eve with God and the crucifix – can also be interpreted in several ways. For example, Goltwurm's reference that Christ was 'put on the cross by our high priests' during the sad years can be connected to the vision of crucifixion. However, the above two images, the eagle and the weeping maid, are sufficient to show that this series of wondrous apparitions is mainly about religious persecutions brought on by the emperor.

'A Big, Wonderful Picture of the Former Elector'

According to Medler's report, after the double-headed eagle vanished, 'a big, wonderful picture of the old Elector Johann Friedrich' appeared. Surprisingly, this was not the only time that the image of the former Elector, who remained in imperial custody until 1552, appeared in the sky of Saxony. Contemporary sources report that people often saw his image during the Interim Crisis. What was the meaning of the vision of the imprisoned prince for people during these stormy years?

Other examples of the vision of Johann Friedrich prove instructive. One year before Flacius's tract appeared, an anonymous pamphlet entitled *Some Visions Seen in the Sky of Holzhausen near Wasserburg ... and Pretin on June 20 of this Year 1548, by Trustworthy People* (*Etliche gesichte sozu Holtzhausen vnther Wasserburg ... Und zu Pretin den xx. Junij. Diese xlviij Jars am Himmel von glaubwirdigen Leuten seint gesehen worden*; 1548) reported that a celestial image of the Elector was seen in two places.[45] According to this tract, a thick black cloud appeared in the sky over Holzhausen one evening, and there came a vision of a man with a black cloth, followed by a knight also in black attire. Then several men, a woman and a knight on a horse came out from the cloud. One man was lying down and the other men tried to revive him, but he did not move. After more men and women came, a large man riding on a fierce-looking horse came out of thick clouds. He was wearing a black hat with a feather and was accompanied by seven retinues. This man looked like 'our former Elector', the author of the text states. The claim of the farmers who saw this sight that they knew Johann Friedrich's face because they had seen him when they were working a construction site in Ingolstadt adds a realistic touch to this narrative. After this group went away, an army of knights and infantrymen came, followed by a camel. Later a cloud came from the south, and there appeared a man with a long coat who looked like Martin Luther. He raised his arm and held a book in his hand. After listing of the names of the five people who saw these strange apparitions, the author recounts another wondrous case from Prettin. First a beautiful deer with big antlers appeared from the moon, followed by two swords with the herald of the Elector; then 'a fat, large man in black, who is like our former Elector, was seen'. Like the case in Holzhausen, a large army of soldiers, knights and servants followed the image of Johann Friedrich.[46] These strange apparitions, which continued for thirty minutes, were witnessed by a pastor and some other people.

Another story of the apparition of Johann Frederick is recorded by both Fincel and Goltwurm. A mysterious incident occurred in Jena on the evening of St Jacob's Day, 1548: around five in the evening, the colour of the sun suddenly turned blood-red, and a large brown *Kugel* (ball) was seen near the sun. After the ball revolved around the sun, the head of a bearded man appeared on the sun and turned his face to the city. As the strange ball turned around the sun again, another head appeared. After an hour 'one more large vision was found over the sun, towards the East, and the Elector of Saxony, Duke Johann Friedrich, much beloved in memory, was truly seen'. After this apparition vanished, a head of a man with a large nose appeared.[47]

In his book published in 1557, Basilius Monner, the legal counsellor of the Elector and former tutor to his lord's two sons, mentions that 'many trustworthy people' saw the *bild* (image) of the former Elector standing in the sky 'frequently'.[48] We do not know where Monner got his information about the visions, but it suggests that news of the Elector's image was widely circulating in Saxony. Thomas Kaufmann interprets these apparitions (in Holzhausen, Prettin and Jena) as 'an expression of the situation of political crisis in Ernestine Saxony and sorrow over the condition of the Elector'.[49] This view seems valid because as we saw in Chapter 1, in many cases celestial apparitions reflected people's anxieties and concerns. For example, according to a Strasbourg broadsheet, a stately funeral procession with mourners and musicians was seen in the sky of Wittenberg in September 1547.[50] If Wittenberg people really saw such an apparition, it could have been a reflection of their concern for the safety of their Elector, who was under a death sentence. However, the above reports of the Elector's apparition suggest a more important factor. Bonner's story of the hero on a white horse shows that the image of Johann Friedrich had almost become a myth.

The most important factor of this myth is that Johann Friedrich had earned iconic status in his subjects' minds through his sufferings for a religious cause. After Johann Friedrich was captured at Mühlberg, the emperor sentenced him to death as a traitor, but the sentence was later commuted to imprisonment for an indefinite period. He was forced to renounce his title as Elector and to surrender a large portion of the territories of Ernestine Saxony, including the Wittenberg region. Although he and his sons were eventually allowed to retain the title of Dukes of Saxony, the size of their territory was largely reduced. Despite heavy pressure and intimidation by the emperor, Johann Friedrich staunchly refused to accept the Augsburg Interim and urged his sons to adhere to orthodox Lutheran faith at any cost and to refuse any compromise over religious matters. He once challenged Charles: 'I stand before your majesty as a poor imprisoned man … I have confessed truth and therefore abandoned possessions, estate, land and people … and have nothing but this imprisoned body … and should I also lose eternal [life] through my recantation? God preserve me from that … It would be wrong if many thousand people would be led into great scandal through my foolish recantation'.[51]

The steadfastness and courage of the captured prince even in the face of execution were strongly contrasted with the passive attitude of Melanchthon and his colleagues in Wittenberg. Gnesio-Lutherans, including Flacius, highly praised Johann Friedrich as a true Christian martyr willing to bear the cross of Christ. As one historian stated, 'in his defeat and imprisonment, the Elector became a transformed, christomimetic figure whose own passion mirrored

Christ's, and whose imprisonment mirrored Daniel in the lion's den'.[52] In a song published in Magdeburg, Johann Friedrich was praised as 'by the Grace of God Elect Martyr of Jesus Christ, Duke of the Afflicted, Prince of the Confessors of the Faith, Heir to Eternal Life. And after this life, Victor over Charles and Judge of his betrayer'.[53] One anonymous text narrates the suffering of the Elector at Mühlberg, deliberately comparing it with Luke's account of the Passion of Christ.[54] The writings of the former Elector became part of the propaganda campaign of the Magdeburgers. His confession of faith presented to the emperor was circulated, and prayers and poetic lines attributed to him and his wife were attached to popular melodies and sung.[55] It seems that Johann Friedrich himself recognized the parallel between his condition and Christ's Passion, and he showed his willingness to follow the way to Calvary. In one of the songs allegedly written by him, the imprisoned prince laments the betrayal of his own cousin and states: 'Now I myself have taken on the likeness of Christ our Lord'.[56]

Kempff, the printer of Flacius's tract, joined this propaganda campaign of the 'Martyr Elector'. Johann Friedrich appears in a large illustrated broadsheet titled *The Difference between the True Christian Religion and False Idolatry Teaching of Antichrist* (*Unterscheid zwischen der waren Religion Christi vnd falschen Abgöttishcen Lehr des Antichrists*) and printed by Kempff, obviously during the Interim Crisis.[57] In this print, the practices of Lutheranism (centred on worship, preaching and Holy Communion) and those of Roman Catholicism (centred on mass, indulgence and blessings) are compared.[58] Friedrich is standing among the audience of Luther's sermon, bearing a big wooden cross on his left shoulder. Here the former Elector is depicted as a faithful martyr who is willing to bear his cross. His pose is odd: though he is showing his back to the viewer, he twists his head backward and looks towards viewers as if he is encouraging them to follow him and bear the cross. The attached anti-papist text was written by none other than Flacius. Though the text does not mention Friedrich, here Flacius and Kempff are collaborating to promote anti-papist sentiment, using the image of the imprisoned prince as a model Christian.

Thus by his sufferings Johann Friedrich became an iconic figure, almost like a saint. It is suggestive that in the Holzhausen case, his image was seen with that of Luther. After his death Luther quickly won saint-like status, especially among Gnesio-Lutherans. The fact that Luther had left many prophetic, pessimistic remarks about the future of Germany, and the apocalyptic disasters of the Schmalkaldic War befell the Church immediately after his death, contributed to his image as the 'Third Elijah', the holy prophet of the End Times. The images of Luther and Johann Friedrich watching their land from the sky reminds us of the iconography of the medieval patron saints protecting the people.

Indeed, Basilius Monner regarded these reports of the celestial image of Johann Friedrich as solid evidence of the former Elector's innocence and the legitimacy of his cause. In 1557 Monner published a treatise which claims that

Johann Friedrich's military campaign against the emperor was legitimate and that he was persecuted because of his faith. In this provocative book, Monner lists various testimonies which show that the former Elector was not a rebel but the innocent victim of an aggressive religious war planned by the emperor.[59] These testimonies include not only those offered by human beings but also those by God, and one of them is the celestial apparitions of Johann Friedrich's image. He states that the fact that many people frequently witnessed the image of the former Elector indicates that Johann Friedrich was neither a rebel nor an evildoer, but a martyr loved by God.[60] Monner also states that many 'wondrous works' ('wunderwercken') occurred during the Siege of Magdeburg. The 'noblest' ('fürnemste') example is that of a mysterious knight riding on a white horse and beating the enemies, frequently seen around the city of Magdeburg during the siege. The source of this report is not certain, but Bonner emphasizes that the image was seen by the besieging army, not by the Magdeburgers. Monner obviously regards the knight as the angel, and the fact that the angel appeared above Magdeburg suggests that the war against Magdeburg was unjust.[61]

Thus, for Monner, the supernatural vision suggests that Johann Friedrich is in God's favour, and obviously he believed that stories of the former Elector's vision had propaganda value. Probably it reflects many people's interpretation of the former Elector's apparition – otherwise an experienced jurist like Monner would not have quoted them as serious 'evidence' for Friedrich's legitimacy. Indeed, he risked his career and probably his own life by publishing this treatise because it enraged Elector August of Albertine Saxony to the extent that Monner would have been severely punished if Duke Johann Friedrich II of Saxony had not sheltered him (Flacius highly praised Monner's faith).[62]

We can now understand why the image of the former Elector was so important in Flacius's tract. It offers another example of God showing his favour to the imprisoned prince, which signifies the legitimacy of Lutheran resistance against the emperor. It also shows what true Christians should do in this time of tribulation: bear the cross and follow Christ, as the pious prince did. At the same time, the image of saintly Johann Friedrich highlights the weak characters of the 'adiaphorists' in Wittenberg.

One interesting point to note is that some accounts of the vision of the prince state that it was a *bild* (picture) of the former Elector, not Johann Friedrich himself. Flacius explicitly states that it was 'a big, wonderful image' ('ein gros hehrlich bild'), and in his account of the Braunschweig case, Fincel notes that it was a picture 'like the one painted by Lucas the painter' ('als hette es Lucas Maler abconterfeiet').[63] The artist Lucas must mean Lucas Cranach the Elder, who painted several portraits of the Elector. The report of the Jena apparition also states that it was like a picture painted by an artist, and Monner states that 'the Elector's picture' ('Churfürsten bilde') was seen. Probably they had to stress that the vision was a painting because Johann Friedrich was still alive, and it could be suspected as a product of magic if a living man appeared in the sky. Thus for Gnesio-Lutherans the appearance of the image of Johann Friedrich is a pure *wunderwerck* of God.

We have examined three types of visions in the Braunschweig *Wunderzeichen*, and it is clear that there is one motif connecting all of these three apparitions: persecution of true believers. Readers of this tract were encouraged to take the side of Christ and choose suffering with the faithful believers, as Johann Friedrich did; God's Judgement was rapidly approaching, so true Christians must prepare for Christ's return by showing their flag clearly.

Melanchthon and Celestial Wonders during the Interim Crisis

In Chapter 2 we saw that Flacius's arch-enemy Melanchthon was greatly interested in astrology and *Wunderzeichen* and had influenced many students at Wittenberg University. What did Melanchthon think about the Braunschweig case? We know that he was not ignorant of the incident; since Medler was his friend, Melanchthon probably heard about it directly from Medler, not through the pamphlet of his ex-colleague Flacius. In his Latin letter to Johann Placotomus of September 1550, Melanchthon mentions the Braunschweig case. He briefly summarizes what the merchants of the city saw in the sky and adds his interpretations.[64] In this comment, Melanchthon focuses on the images of the camel and Eve and interprets them as the signs of the Turk's threat to Germany ('bella Turcicia') and the difficult conditions of the Church ('mediocrem Ecclesiae statum'). The prophecy of Johann Hilten, a Franciscan monk in fifteenth-century Eisenach who reportedly prophesied that the Turks would dominate Germany and Italy in the future, is mentioned to support Melanchthon's views.[65] However, he does not add any comment about the images of an eagle and a weeping maid. This does not mean that Melanchthon was not interested in these visions; it seems that he discussed the Braunschweig case in his university lectures long after the Interim Crisis.

This story appears in a book written by Leonhard Kentzheim, the Superintendent of Liegnitz. In this book, published in 1583, Krentzheim warns readers that the world is approaching its end and a number of signs are appearing in the sky. To buttress his view, Krentzheim states that Melanchthon himself told an interesting story of portents. During his lecture on Luke 21 in 1559, attended by one hundred students, Melanchthon told the students:

> You young people are still living in the golden time, but soon a dreadful, sad time will come as wonder-signs appear and are witnessed everyday. Yet due to their great confidence and stubborn evil nature, people ignore these signs, and [these warnings] are all in vain.[66]

Then Melanchthon turned to the Braunschweig case as an example of such warning signs. He explained each apparition one by one and sometimes added his own interpretation. First he mentioned the three moons, then a fiery lion, and then 'a wounded eagle, which meant the death of Emperor Charles V'. The image of Johann Friedrich meant that Friedrich would recover his land and his people. Then Melanchthon added an apparition that is not described in Flacius's pamphlet: an image of Luther, which meant that God built his Church through the preaching of the Gospel (did Melanchthon mix the Braunschweig case with the Holzhausen case?). He mentions the image of Adam and Eve but does not add any comment. The next image he describes is a city on fire with a camel standing nearby. Flacius's pamphlet does not say that the city was in flames, and Melanchthon interprets this as a sign of the invasion of the Turks. Here he quotes Johann Hilten's prophecy that by the year 1600 Gog and Magog would invade Germany and Italy and dominate them. Christ on the cross, a group of soldiers with spears and a gathering of the apostles all mean that the Christian Church is persecuted but will be preserved. The last apparition, a man with a sword and a weeping maid, indicate the end of the world.[67]

Robin Bruce Barnes mentions this episode in his book, but he states that he could not verify it in either the *Corpus Reformatorum* or the *Supplementa Melanchtoniana* and 'these particular words may be inexact or even apocryphal', although they are 'certainly not uncharacteristic of Melanchthon'.[68] However, Krentzheim's narrative should not be discounted. We know that Melanchthon was greatly interested in *Wunderzeichen* and often exchanged information about prodigies through correspondence. Indeed, Melanchthon does describe the Braunschweig case in his letter of 1550, and there he also mentions Johann Hilten's prophecy about the Turks. Probably the episode was not recorded in his official writings because Melanchthon told the story as leisurely talk during one of his lectures.

Melanchthon's interpretation of the 'wounded eagle' as a sign of the death of Charles V seems strange because the emperor did not die until 1558, ten years after the apparitions. However, possibly this interpretation was a reflection of wishful thinking which he kept in his heart during the Interim Crisis. Several sources suggest that Melanchthon indeed tried to predict the death of Charles based on his observation of stars and hoped that the present political crisis would come to an end soon. According to Flacius's book published in 1560, Melanchthon observed the eclipse of the moon in April 1548 and predicted that Charles would die in August.[69] When the emperor survived the month, Melanchthon consulted Ptolemy's book and changed the date of Charles's passing to August 1549.[70] Flacius called such speculations 'godless expectation' ('Gottlose hoffnung') and ridiculed Melanchthon's trust in the stars. Placing one's hope of deliverance on the stars and the emperor's death was nothing but foolish.[71] For Flacius, what was needed now was a sincere confession of faith, not star-gazing.

Although this account is from Melanchthon's enemy's camp, historians generally accept it at face value.[72] Since Flacius was still in Wittenberg in 1548, it is not strange that he witnessed the Preceptor placing his hope in the signs that he saw in the sky. Such an attitude seemed to be meaningless for Flacius, but Melanchthon was serious about the results of astrological calculations, as the following story shows. In 1548 Melanchthon received an invitation from Thomas Crammer to assist the church reform in England under Edward VI. One of the reasons why Melanchthon declined this timely invitation (which offered him an opportunity to escape from the turmoil in Saxony honourably) is that when he was a child, he was warned by the prominent astrologer Johann Virdung that he would be shipwrecked in the Baltic someday.[73] Considering these factors, it is not difficult to imagine that Melanchthon was encouraged by the report of the Braunschweig case. Maybe he thought that the eagle was dying for his wound – one believes what he wants to believe.

Even if we discard Flacius's testimony as unreliable, it is an undeniable fact that Melanchthon mentions the prospect of Charles's death during the time of the eclipse in his letter to the King of Denmark Christian III (dated 25 May 1551). In the epistle, he bitterly laments the condition of Germany: 'The sorrow in Germany is so great that men can clearly understand that Christ's Word indicates no other than this grief: There will be anxiety in the End Time, and people will suffer. And confusion in Germany has no end'.[74] He recounts some recent news at the end of the letter, including a brief report of the emperor's illness: 'On 20 February, during the time of the eclipse, the emperor was in very serious condition, and there was not much hope for his survival. Nonetheless after several days he became better'.[75] He also attached news of the *Wunderzeichen* that appeared near Torgau in the previous year: the sun became black, broke in two parts, and fell. After this, people saw a great battle fought in the sky, and there appeared a deer standing on a mountain.[76] Melanchthon interpreted the deer as a symbol of the Church which God sustains in spite of war. Thus Melanchthon's expectation of the emperor's death was frustrated, but he encouraged himself by interpreting the Torgau *Wunderzeichen* as a sign of God's protection of the Church in political turmoil. In his article on Melanchthon's character, Markus Wriedt points out that there is a marked tension between *Angst* and confidence in God throughout his life (Wriedt uses the word *Angst* to express the 'comprehensive sense of responsibility', but in Melanchthon's case it is closer to 'worry which come from a sense of responsibility'). This tendency can be applied to his belief in astrology:

> In connection with astrology, we again see Melanchthon's posture between Angst and confidence. On one hand, he continually worries about natural happenings being dark omens. On the other hand, he steadfastly trusts in God's saving words of the Gospel and knows he is sheltered within the Creator's plans for salvation.

This suggests that Melanchthon's optimistic interpretations of these wondrous signs can be regarded as a part of his pious trust in God, but surely Flacius had a different opinion.[77]

These examples show that Melanchthon was truly looking for a sign of deliverance from tribulation in the stars and *Wunderzeichen*. Probably he reacted in the same way when he heard the report of the Braunschweig case, which had to disappoint Flacius deeply. The fact that Flacius exposed Melanchthon's stargazing during the years of crisis twelve years after the incident suggests the depth of his contempt – he was clearly aiming at character assassination. When he published the tract on the Braunschweig case, Flacius certainly knew his former teacher's love of stories of wonder. Did Flacius try to ridicule his former teacher by deliberately using news of *Wunderzeichen* for the promotion of his anti-Interim message? Considering Flacius's obsessive character and the fact that Flacius reprinted Melanchthon's account of the Papal Ass with biting remarks about the author, this may be more than a remote possibility.

Conclusion

Thus the Braunschweig case contains a series of messages that represent the core of the Magdeburg theologians' world view: the weeping maid symbolizes the Church in danger; the eagle signifies the wrong course taken by the emperor; and the image of the imprisoned Elector shows that people should follow in his steps by bearing the cross. The passage from Luke 21 printed on the title page, 'when these things begin to take place, straighten up and lift up your heads, because your redemption is drawing near', is an appropriate quotation. People saw not only the wondrous celestial signs but also the various tribulations foretold in Scripture, symbolized in these apparitions, becoming real. The whole series of apparitions is like an emblem of the End Times.

Why did Flacius decide to use a *Wunderzeichen* story as his propaganda tool? There are two possible reasons: first, Flacius obviously knew of the popularity of *Wunderzeichen* stories among people, especially during times of difficulty. In his wonder book published within a decade after the Interim Crisis, Fincel recorded various stories of extraordinary phenomena observed during the years following the Schmalkaldic War. Since Flacius mentions the appearance of mysterious springs of blood in his pamphlet, most likely he had heard some of these stories of wonder, such as the ominous blood-red sun seen in Weimar around the time of the Battle of Mühlberg.[78] Probably he heard the news of the explosion of the gunpowder-magazine in Mechelen caused by a thunderbolt. It was reported that the magazine had stored gunpowder and weapons intended to be used for the war against the German Protestants, and this incident was regarded as a manifest sign of God's displeasure with the emperor's military campaign.[79] Likewise, people talked about a church in Meissen destroyed by lightning, when people were celebrating the defeat of Johann Friedrich with the singing of 'Te Deum'.[80] These stories indicate that many people desired to find divine signs against the enemies of the Protestants. As one of the leaders of the anti-Interim campaign, Flacius believed that by using *Wunderzeichen* news he could reach a broader range of readers. People's familiarity with Christ's apocalyptic prophecy and its connection with *Wunderzeichen* made the story of the Braunschweig apparitions more appealing.

Since at least four editions of this pamphlet were published, obviously Flacius expected that it would be circulated among a mass audience. The content of Flacius's preface, which explains the basic doctrines of the Lutheran faith and its difference from those of the papists, also suggests that the tract was intended for a wide readership, probably for common people who could not explain the difference of doctrine by themselves or for those wavering between the two faiths. Flacius could expect to use this *Wunderzeichen* report as an opportunity to teach readers the summary of the Lutheran faith and to urge them to stand firm against the papists. Of course he did not forget to castigate the *gesellen* of Antichrist, though he did not use the term 'adiaphorists' itself. His avoidance of the theological term in this tract may reflect his consideration of the level of readers' theological knowledge.

Another possible reason for the use of the *Wunderzeichen* story is the desire to avoid a direct attack on the emperor. While Magdeburg theologians bombarded the pope, Catholic bishops and Wittenberg theologians with their accusations and personal attacks, they knew that it was not wise to accuse Charles directly. Such action could alienate many Lutherans who were struggling to find a balance between their loyalty to the emperor and their Lutheran faith. The Senate of Magdeburg repeatedly declared that they were willing to subject themselves to the emperor's authority if their religious liberty was guaranteed, and this became the official stance of the city during the crisis. Consequently, when they wanted to criticize Charles's policy, the Magdeburg theologians had to devise special means. One of the strategies was to present materials which presented Charles in negative light while leaving readers to interpret the texts by themselves.

One example is Flacius's publication of Charles's notorious edict against the Protestants in the Netherlands in 1550. According to the preface, written by Flacius, what prompted him to publish this edict was the opinion of Johann Agricola, a court preacher of Brandenburg who contributed to the formulation of the Augsburg Interim. Agricola was claiming that the emperor was in fact a Lutheran at heart, and Flacius expressed his serious concern that such 'a dangerous lie' could mislead people and give them the vain hope that the emperor's Interim was not as bad as it was said.[81] Readers were urged to examine the edict carefully to see whether Charles, who ordered bloody suppression of Protestantism in the Low Countries ('Very trustworthy and sensible people say that 70,000 people were gruesomely killed in the Netherlands for religious reasons'[82]), could really be called a Lutheran. Flacius's *Wunderzeichen* tract could be a product of this kind of strategic manoeuvre. If the tract was intended for a wide circulation, Flacius had to avoid a direct attack on the emperor, which could be regarded as seditious and offend many readers. The image of the self-hurting eagle is convenient for Flacius's purpose, because no one would miss what was symbolized by the double-headed eagle while the name of the emperor was not directly mentioned. And it was believed that celestial signs were sent by God and reflected divine will. Considering these factors, it is understandable that Flacius incorporated this *Wunderzeichen* report into his propaganda campaign.

This chapter has examined how stories of wonder were used for political and polemical purposes during the time of crisis. The popularity of *Wunderzeichen* made them valuable tools to capture people's attention. It also gave the author an opportunity to insert messages he intended to disseminate. Like prophetic literature in the sixteenth century, the enigmatic images of apparitions were convenient to carry political implications which could not be discussed openly. Melanchthon's behaviour shows that it was not unusual to distil political messages from celestial signs, and Monner's quotation of stories of Johann Friedrich's apparition in defence of his former lord suggests that people were serious about the political implications of celestial apparitions. Flacius mocked Melanchthon's star-gazing as a sign of his weak faith, but he himself employed *Wunderzeichen* in his propaganda campaign to influence as many people as possible. Thus Flacius's tract and the world surrounding it both show the dynamism of *Wunderzeichen* as media wherein political, theological and psychological discourses interact.

Since the 1549 tract was Flacius's only writing on wonders, his interest in this topic might be a limited one. However, Flacius had a friend who held a passionate interest in the world of wonders and tirelessly collected *Wunderzeichen* stories even during his difficult years of exile. In the next two chapters, we shall examine the writings of Christoph Irenaeus, a passionate adherent to Flacius's theology on original sin. Like Flacius, he used the reports about wondrous phenomena to attract people's attentions and to discredit his theological opponents. At first Irenaeus aimed his attacks at the theologians whose doctrinal understandings fell from the mainstream of Lutheran faith, but later his target became the very centre of his own church: the Formula of Concord, which became the official confession of the Lutheran Church.

4 IRENAEUS AGAINST 'SPIRITUAL WOLVES': POLEMICAL USE OF *WUNDERZEICHEN*, I

The end of the Interim Crisis did not bring peace to the Lutheran Church. The mutual suspicion of Philippists and Gnesio-Lutherans fostered a number of bitter theological controversies between them, and doctrinal disunity pestered the Church for the next three decades.[1] The protagonist of the following two chapters is Christoph Irenaeus, who was deeply involved in these theological polemics. One interesting fact about Irenaeus is that he was greatly interested in wonders in his days, diligently collecting tales of wonder throughout his life. In his four books about *Wunderzeichen*, he catalogued hundreds of stories about comets, *Fewerzeichen*, monstrous births, natural disasters and other shocking incidents. However, he did not collect these tales to amuse readers; his goal was to use these stories to promote his theological agendas and discredit his opponents as *Schwärmer* ('enthusiasts'/mad). How did he connect the stories of wondrous phenomena with theological controversy? What kind of relationship did he perceive between *Wunderzeichen* and the doctrinal disunity of the Lutheran Church?

The name of Irenaeus is hardly a familiar one, even for Reformation historians. Before Philip M. Soergel devoted a chapter in his *Miracles and the Protestant Imagination* to Irenaeus's wonder book, the only English study on Irenaeus's theology is Robert Kolb's article that discusses Irenaeus's role in the polemics against Jakob Andreae and the Formula of Concord.[2] Robert Christman's *Doctrinal Controversy and Lay Religiosity in Late Reformation Germany* (2010) mentions Irenaeus's active involvement in the Flacian Controversy in the Mansfeld region and offers valuable background information on Irenaeus's polemical works;[3] yet since the focus of this work is not on Irenaeus but on his friend Cyriakus Spangenberg, Irenaeus remains in the background. Even in German studies, the number of works that discuss Irenaeus's writings is limited. Irene Dingel's extensive study of the dissenting voices against the Formula of Concord carefully explains the contents of some of Irenaeus's anti-Concord works in the 1570s and the 1580s, together with the responses of his theological opponents.[4] However, understandably, his works in the 1560s are not discussed because they do not relate to the scope of Dingel's study.

Irenaeus appears in several other studies in connection with Lutheran apocalypticism and wonder books. Barnes mentions him in *Prophecy and Gnosis* as an author of a book on comets.[5] Rudolf Schenda explains the essence of Irenaeus's three books on prodigies in his article about the history of German *Prodigiensammlungen* and highly praises his creative style of writing.[6] Schenda also wrote a three-page article about Irenaeus for *The Encyclopedia of Märchen* because Irenaeus's books contain sixteenth-century versions of famous legends such as 'Hatto, the Cruel Bishop of Mainz', 'The Pied Piper of Hamelin' and 'Merlin the Wizard', offering interesting materials for today's fairy-tale research.[7] In his massive study on the Lutheran apocalyptic literature, Volker Leppin refers to Irenaeus's two books on wonders and recognizes the polemical natures of these works. He briefly explains that Irenaeus's comet book was written as a criticism of the Formula of Concord, especially its doctrine of original sin, and the Formula's chief architect, Andreae.[8] Dingel also mentions Irenaeus's comet book in her study and points out that Irenaeus connected his eschatological conviction with natural phenomena and his theological opponent.[9] What is desired now is a study that combines these findings of past scholarship and systematically examines the theological and polemical agendas of Irenaeus's *Wunderzeichen* works.

This step has been taken by Soergel's insightful chapter in his *Miracles and the Protestant Imagination*.[10] Describing the outstanding features of Irenaeus's three wonder books, Soergel expresses his fascination with the polemist's creative reading of natural phenomena and wonders as icons of divine will, in addition to his picturesque writing style with rich vocabulary use. What is revealed is Irenaeus's 'penchant for conducting a mildly ironic, yet ultimately good-natured argument with his readers; and a fluent, animated style that dispensed grim spiritual medicine while still managing to entertain'.[11] His (and also his friend Spangenberg's) passion to catalogue all kinds of individual vices reflects the 'often highly scholastic, hairsplitting morality' of Gnesio-Lutherans, and it shows their desire to 'deepen their audience's awareness of the interconnections between every human desire, thought, and action – to consequently stir greater introspection and self-discipline'.[12] The polemical intentions of Irenaeus's books on comets and monstrous births and their connection with Flacius's theology are also discussed. However, Soergel does not forget to mention that what is at stake for Irenaeus was not only his opposition to the doctrinal stance of the Formula but also his criticism of mixing natural philosophical perspectives with Christian faith. For Irenaeus, the interpretation of postlapsarian humankind presented by his theological opponents is solely based on their philosophical argument on the accidental nature of sin, and it treats the horrific effect of the Fall too lightly. Reflecting Luther's view of the degenerated condition of the earth as divine curse, he regards the presence of hideous creatures such as vermin and reptiles on the earth as the manifest sign of God's condemnation of human sin. This perspective later developed into his unique interpretation of monstrous races as emblems of the monstrosity of human nature after the Fall. Soergel concludes his analysis by pointing out Irenaeus's 'ambiguous, puzzling, and troubling' view of human nature expressed in his book on misbirths, which can be compared with the mysterious images in Giuseppe Arcimboldo's paintings.

Based on these past studies, this and the following chapter explore Irenaeus's four books (including two works produced in the early 1560s that have not yet received sufficient scholarly attention) and analyse the development of his theological focus and its interconnection with his interest in *Wunderzeichen*. One point that should be examined carefully is Irenaeus's strategic use of wonder stories to catch readers' attention and his ability to transform these stories into polemical weapons. For him, *Wunderzeichen* were nothing but manifest signs of God's wrath (*Zornzeichen*) against 'false preachers' and their followers. By describing various examples of *Wunderzeichen*, he intended to present proof that God was angry with the enemies of the Gospel – those who disagreed with his theological views. Thus Irenaeus's interest in wonders, theology and polemical intention cannot be separated. What is revealed is the dramatic life of a theologian who had a remarkable talent for writing and an intense curiosity, but who did not hesitate to sacrifice his career for his theological convictions – a number of times.

Not all of his wonder books discuss celestial phenomena. The second book is mainly about watery disasters such as floods and storms, while the fourth book focuses on monstrous births. However, the contents of all four books must be viewed for three reasons: first, in sixteenth-century society both natural disasters and misbirths were generally regarded as parts of *Wunderzeichen* and also interpreted as signs of divine wrath against human sinfulness. In addition, by examining all four books, we can see how Irenaeus developed his skill of using stories of wonder to attract readers and lead them to his polemics. Finally, Irenaeus's connection of monstrous births with his understanding on human nature, described in the fourth book, is a logical conclusion of his theology on original sin, which permeates all of the four works.

Since his works are still largely unknown to most scholars, this study presents the contents of Irenaeus's books section by section. In this way, we can comprehend the flow of his arguments, his thinking patterns and his creative skill in guiding readers. In this chapter, his three early books – *Apostles' Creed* (*Symbolvm Apostolicvm*; 1562–3), *Print of a Dreadful Sign of Wrath* (*Abdruck Eines schrecklichen Zornzeichens*; 1564) and *Mirror of Water* (*Wasser-Spiegel*, 1566) – are explored. His earliest publication, *Symbolvm Apostolicvm*, contains an extensive criticism of astrology, and there he already shows his strong interest in celestial phenomena. In the second book, *Abdruck*, Irenaeus took the style of an Old Testament prophet and quotes almost one hundred passages from Scripture to urge people to repent and abandon their sinful lives. He threatens readers that

God will soon send terrible punishments on Germany since people do not take heed of godly preachers but follow the errors of 'false teachers'. One interesting element of this work is that Irenaeus inserts his own vivid account of *Wunderzeichen* in the sky over Eisleben in 1564 (mentioned in Chapter 2) at the start of the book; this experience obviously inspired him to write the work. In his next publication, *Wasser-Spiegel*, Irenaeus catalogues hundreds of wondrous stories relating to water. While the *Abdruck* accuses false teachers and unrepentant sinners in general, in this book he specifies Calvinists, Synergists and others as the false teachers of his day and uses stories of wondrous incidents to condemn their teachings. The following chapter will analyse the relationship between Irenaeus's wonder books in the latter half of his life and his involvement in the Flacian Controversy over original sin. The main sources of this chapter are Irenaeus's *Prognosticon from God's Word* (*Prognosticon Aus Gottes Wort*; 1578), *Exposition of the Gospel Luke 21* (*Erklerung des Euangelij Luce XXI*; 1581) and *The Monsters* (*De Monstris*; 1584). His *Image and Mirror of Mankind* (1582) will also be mentioned briefly as a reference. Throughout these books, Irenaeus attacks the same target – Jacob Andreae and his Formula of Concord. What provoked Irenaeus's wrath is his opponent's theological understanding of original sin as incorporated in the Formula, and Irenaeus used the above works to attack the Concord project itself.

The Life of Irenaeus and the Formulation of the Concord

Irenaeus was born in Schweidniz in Silesia (now in Poland) in 1522.[13] After attending the school of Valentin Trotzdorf in Goldberg, he went to Wittenberg in 1544 and studied under Melanchthon. Between 1545 and 1547 he served as the school rector of Bernburg and later took up a position at a school in Aschersleben. In 1549 he received a Master's degree from Wittenberg University. His marriage in 1551 to Irene Plateanus, the daughter of the Superintendent of Aschersleben, Petrus Plateanus, shows that young Irenaeus was steadily building his career as city clergy. Ordained by Johannes Bugenhagen in February 1552,[14] he became deacon of Aschersleben. After becoming archdeacon of the city in 1559, he was appointed as court preacher of Weimar. However, when he was serving at the court, the Synergist Controversy over the role of the human will started between the two ducal theologians, Viktorin Strigel and Flacius. Strigel argued that despite the Fall, human will retained its power to cooperate with the work of God's grace, while Flacius completely rejected such a claim. The hero of Magdeburg insisted that one had no power to contribute to his or her salvation because the Fall totally corrupted the very substance of human nature. Soon Flacius's group fell out of grace with their lord, Duke Johann Friedrich II of Saxony, and Strigel's opinion gained the upper hand. Because he refused to accept Strigel's view, Irenaeus had to leave his post in 1562.

He then took the post of pastor at St Peter and Paul's Church in Eisleben, where Luther had been baptized. It is not difficult to imagine that Irenaeus was proud of a post with such memorable history. In the preface of his first publication, under his name, he adds the title 'The pastor of St. Peter and Paul in Eisleben, the church where Dr. Martin Luther was baptized on Martin's Day, 1483'.[15] In 1562–3 he published the *Symbolvm Apostolicvm*, and his friend Cyriakus Spangenberg warmly welcomed his colleague's first publication by contributing a lengthy preface to the book. In 1566 Irenaeus nearly became a victim of the rampant pestilence that claimed many lives in the region, but he managed to recover.[16] In the same year, Duke Johann Wilhelm of Saxony called him to Coburg as his court preacher. After Duke Johann Friedrich II was arrested by the emperor because of his Gotha Rebellion and lost his title, Duke Johann Wilhelm succeeded his brother and took Irenaeus to Weimar in 1568. Thus Irenaeus triumphantly returned to the post of ducal court preacher. Using his influence at the ducal court, Irenaeus worked to support the placement of Flacian theologians in church offices and at the University of Jena. When representatives of the Gnesio-Lutheran party and the Philippists met at Altenberg in 1568–9 to resolve their differences over justification, free will and good works, Irenaeus attended the sessions. Unfortunately, the colloquy failed to reach any productive agreement. In 1570 he participated in the visitation of churches in Franconia.

Up to this point his ecclesiastical career had been promising, but soon stormy winds started gathering against him. Fearing that theological disunity would weaken the Lutheran Church and make it vulnerable to Counter-Reformation influences, Lutheran princes were pressing theologians to restore the unity of the Church. Andreae, a theologian of Tübingen, received a commission from Duke Christoph of Württemberg to work for the Lutheran Concord. Andreae prepared a brief, simple statement concerning the five controversial doctrinal issues and urged other theologians to subscribe to it. There was no condemnation of theologians who taught 'erroneous' doctrines. In 1569, encouraged by support from several of the princes, Andreae launched his travel to all over the Empire to call for subscription. However, Andreae's efforts met with strong opposition in Saxony. The Gnesio-Lutheran theologians of Jena University rejected his plan and warned Andreae that unless all errors were specified and condemned, an agreement would cause further disunity.[17] Irenaeus immediately joined the chorus of criticism and published a pamphlet titled *Warning and Causes* (*Warnung vnd vrsachen*; 1569). In this work, Irenaeus castigates those who want to grant 'Amnestia' (amnesty) and insists that there can be no reconciliation unless false teachers recant their errors. Even Augustine and Luther made errors, but they later admitted their mistakes publicly and retracted their previous views.[18] Those who want to give 'Amnestia' to erroneous teachers act like a 'new Tetzel' by wishing to erase the sins of *Schwärmer* with indulgences. The pamphlet clearly shows Irenaeus's desire for purity of doctrine and aversion to any political solution for theological controversies. When Andreae was staying in Weimar to negotiate with the duke, Irenaeus attacked Andreae's project from the pulpit.[19] Thus Andreae left Saxony without success, lamenting that the Weimar theologians possessed neither the Holy Spirit not common sense.[20]

Soon, however, Irenaues's own position in the ducal court became unstable. In 1570 he was transferred to Neustadt an der Orla as the Superintendent, probably due to his stance on the Flacian Controversy. The origin of this controversy is Flacius's view that original sin had become the substance of human nature through the Fall (the details of the controversy will be examined in Chapter 5). Irenaeus passionately defended Flacius's position, while most of the leading theologians, including Irenaeus's friends at Jena University, eventually took the opposite side. Soon he found that it was no longer possible to retain his present post. In January 1572 Irenaeus submitted his *Confession of these Two Propositions or Debate* (*Bekentniß von diesen zweyen Proposition oder Redent*) to the Consistory of Jena with three other theologians and criticized his opponent's understanding of the issues.[21] In the same year, Johann Wigand, a prominent professor of Jena, replied to Irenaeus's challenge with a lengthy book titled *Concerning the Patchwork of Master Irenaeus* (*Vom Flickwerck M. Jrenaej*; 1572) and rejected Irenaeus's argument as 'Manichean enthusiasm' ('Manichaeer Schwermerey').[22]

Fortunately, Irenaeus was not helpless. In 1572 Count Volrad of Tal Mansfeld, a staunch adherent of the Flacian view, called Irenaeus to be a preacher at the Mansfeld fortress. During the next two years, together with his old friend Spangenberg, Irenaeus actively worked for the Mansfeld theologians' campaign against their opponents. However, his days in Tal Mansfeld came to an abrupt end on 31 December 1574, when Mansfeld's feudal lords, Elector August I of Saxony and Joachim Friedrich of Brandenburg, administrator of the archdiocese of Magdeburg, intervened to suppress the Flacian campaign by force. They sent five hundred soldiers to Tal Mansfeld, expelled the city's pastors, and arrested a number of citizens.[23] Irenaeus, his patron, Count Volrad, and Spangenberg had to go into exile. This was the beginning of Irenaeus's long life in exile. Before 1590 he experienced banishment seven times.[24]

It seems that Irenaeus temporarily returned to his hometown of Schweidniz, probably because some Silesian Lutherans were still favourable towards the Flacian position at this time;[25] but he could not remain there for long. Then he seems to have resided in Hesse and the Lower Rhine region for some years.[26] In 1577 he strengthened his bond with Flacius by forming a family union – his son Wolfgang married Anna, Flacius's daughter.[27] Between 1579 and 1582 he lived in Buchenbach, where his wife, Irene, died in 1582.[28] He was afterwards banished from there, but was able to obtain a place to stay in Nuremberg, thanks to one of its citizens, Thomas Itzenberg.[29] Like many other Flacian exiles, he later moved to Lower Austria and served as a pastor at Horn until 1585. When a colloquium was held at Horn in 1582 and a confession upholding the Flacian position was compiled, Irenaeus added his signature.[30] There he actively participated in the exiled theologians' campaign against the Formula of Concord. After being expelled again, Irenaeus returned to Buchenbach and died there in 1595.

What frustrated Irenaeus most in this exile period was the success of his old foe, Andreae. In 1573 Andreae publicized a new proposal for a Concord, entitled *Six Christian Sermons on the Divisions*. In this text, Andreae defines which views are orthodox Lutheran interpretations and specifies who the erroneous teachers are.[31] Flacius's interpretation of original sin is rejected, together with the teachings of Osiander, Major, Strigel and the Crypto-Calvinists. Encouraged by the favourable responses of other theologians, Andreae compiled confessional articles based on this text, and Martin Chemnitz and David Chytreaus were commissioned to revise it. By their efforts the Swabian-Saxon Concord (1575) was formed. Next year, commissioned by Elector August of Saxony, Andreae, Nicholas Selnecker, Chemnitz, Chytraeus and Andreas Musculus completed the Torgau Book. Its revised version, the Bergen Book (1577), became the definitive text of the Formula.[32] By 1580 the Formula had been subscribed to by three electors, twenty dukes and princes, thirty-five imperial cities and 8,188 clergy.[33] In the same year, the Formula and other essential confessions were published as the Book of Concord, and Andreae finally achieved his dream of Lutheran doctrinal unity.

For Irenaeus, this was nothing but a nightmare. A confession that defines original sin not as the substance of human nature was, for him, not only unacceptable but blasphemous. The fact that it was adopted as the official confession of the Lutheran Church convinced Irenaeus that the whole world had fallen into apostasy – truly an apocalyptic development. Throughout the rest of his life, Irenaeus poured his energy into a massive number of writings against the Concord and its doctrine of original sin.[34] These books annoyed the formulators of the Concord so much that Chemnitz and other theologians compiled a voluminous refutation titled *Refutation of Irenaeus* (*Refutatio Irenaej*; 1583) to defend the Formula's position on this issue.[35]

Symbolvm Apostolicvm (1562–3)

Irenaeus's first publication was *Symbolvm Apostolicvm: Das ist Die Artickel vnsers Christlichen Glaubens ausgelegt.* The purpose of this book was to explain the articles of true Christian faith and to refute all errors and heretical opinions from ancient to contemporary times. He opens the book with a quotation from the prophecy of Simeon: 'this Child is appointed for the fall and rise of many in Israel, and for a sign to be opposed' (Luke 2:34–5). It is true, Irenaeus states, that as Matthew 21 shows, Christ is a stone that was thrown away by the builders – the high clergy and potentates of this world, such as popes, emperors, kings, princes and other mighty people.[36] As Simeon predicted, true believers stand on Christ, while false believers stumble over him and fall. Simeon also predicted the suffering of Christ, and this prophecy can be applied to true believers as well. From the beginning of history, false believers have opposed true believers: Cain and Abel, Ismael and Isaac, Esau and Jakob, pagan emperors and Christ's Church, and papists and Lutherans. Today the true teachings of Christ are under attack from Calvinists, Anabaptists, 'Stenkfeldisten' (a derogatory name for the followers of Caspar Schwenckfeld), Antinomians, Intermists, Adiaphorists, Majorists, Osianderists and Synergists.[37] Beware, for wherever God's Word is taught properly and true worship is observed, Satan spreads false teachings, false worship and idolatry, and we must always remain alert because the devil disguises himself as an angel of light.[38] This introduction can be called the mission statement of Irenaeus's whole life as a polemist, and the same conviction is repeatedly presented in his subsequent writings.

The *Symbolvm Apostolicvm* is divided into three parts. In the first, Irenaeus explains the content of the Apostles' Creed and examines the various theological errors being practised against it. The second part is about the process of salvation, followed by the third part about the work of the Holy Spirit. Here we focus on the first part of the work, where Irenaeus refutes astrology. This section contains several interesting points. First, looking at his argument about the stars, we can comprehend how he interpreted celestial phenomena. Second, Irenaeus shows his interest in *Wunderzeichen* as a sign of God's wrath, and this suggests that from the very start of his writing career, he was interested in stories of wonder. Third, his arguments show his strong emphasis on the power of original sin and the depravity of human nature. This theological position dominates Irenaeus's writings throughout the rest of his life. We see that Irenaeus's two topics of lifelong interest – *Wunderzeichen* and original sin – are already discussed side by side in his very first publication.

In the section 'Concerning Pagan Superstition and the Abuse of Astrology', Irenaeus states that the belief that the movements of stars can influence human inclinations to commit sin and create social disorder is nothing but superstition against God's Word. The Bible tells us that there are only three primary causes of human sin: the original sin inherited from Adam, the temptation of the devil and the evil examples of the world.[39] Stars cause no one to sin because the human mind is ruled by either God or the devil.[40] Irenaeus quotes Luther's comment on astrology in his *Lectures on Genesis*, which states that astrology cannot be called a science because it lacks clear proof.[41] Some people insist that astrology should not be disregarded because astrological speculations sometimes predict important events of rulers and lands, but such speculations are fantasy brought on by Satan, as said by Augustine.[42]

This does not mean that stars have no role given by God. The section 'What is the Usefulness, Power, and Work of Stars?' is significant because here Irenaeus expresses his interest in *Wunderzeichen* for the first time. The primary role of stars is, as Genesis states, that the sun and the moon rule day and night. Stars also serve as signs which show that there is one God who created them. When eclipses and other phenomena appear, they serve as signs of God's wrath (*Zornzeichen*), warning people of God's punishment and urging them to repent. After quoting Luther's remark on celestial signs that are supposed to appear before the end of the world, Irenaeus mentions the case of the strange sign that appeared in the sun shortly before the capture of Elector Johann Friedrich, followed by the Interim Crisis.

Another function of the stars, he states, is that the sun and the moon are symbols of the relationship between Christ and his Church: as the moon shines, reflecting the sun's light, so the Church shines, reflecting Christ's glory.[43] At the end of this section, Irenaeus suddenly returns to his explanation of *Zornzeichen* and describes various types of signs of God's wrath: 'However, when he punishes, he shows some special signs of wrath in the sky, such as: *Chasmata* and fire in the sky; *Parelij* [parhelia], many suns [seen] at the same time; comets; terrible weather on the earth, storms, and such thing'. As an example, he mentions the dreadful *Zornzeichen* that appeared on the Day of the Holy Innocents in 1561.[44] An ancient saying is quoted after this paragraph: 'When men see a comet in heaven, punishment and disaster will occur'.[45]

The uniqueness of Irenaeus's view is his emphasis on the utter depravity of human nature as the grounds for his argument against astrology. There are several examples of anti-astrological literature in the mid-sixteenth century, such as Jean Calvin's *Contre l'astrologie* (1549) and Johann Stoltz's refutation of Hebenstreit's *Practica* (1554). Since Stoltz was Irenaeus's predecessor in Weimar, Irenaeus may have read this work, but none of these works uses original sin as the base of their arguments.[46] Original sin is the very core and foundation of Irenaeus's theology, his whole world view. Indeed, immediately after this section on astrology, Irenaeus turns his argument to doctrinal errors concerning original sin. He laments that even in the Christian Church, the doctrine of original sin had been distorted by Pelagians, papists and 'free-will' theologians.[47] In conclusion, he emphasizes that it is clear from the Bible that human nature is damned and has become death itself through the Fall, and this truth is supported by the words of the Holy Fathers, Paul, Christ and Luther.

Thus Irenaeus shows a strong adherence to the Flacian stance on human nature from the very beginning of his career as a polemicist. The controversy over original sin escalated to critical levels in the late 1560s, but Irenaeus's argument in the *Symbolvm Apostolicvm* shows that he was interested in this topic from an earlier point. However, at this moment he had yet not connected outright doctrinal points with *Wunderzeichen*, as he does later. In the *Symbolvm*

Apostolicvm, *Wunderzeichen* are only mentioned as a sign of God's wrath. This is a conventional interpretation, but it becomes an important topic in his writings for the rest of his life. Within two years, and triggered by a real event, he published a new book that again focused on this subject.

Abdruck Eines schrecklichen Zornzeichen (1564)

Irenaeus's *Abdruck Eines schrecklichen Zornzeichen* was published by Urban Gaubisch in 1564. This book can be divided into two parts: the first consists of a preface written by Hieronymus Mencel and a narrative on shocking celestial phenomena that appeared in the sky over Eisleben; the second part, 'A Christian and Necessary Admonition to this Evil Last Age Based on God's Word', is an exposition on human sinfulness, God's wrath, examples of God's *Zornzeichen*, the danger of false prophets, and the signs of the imminent coming of the Last Judgement. The overall message of this part can be summarized as 'repent or disaster!', like the *Wunderzeichen* discourses by other authors. However, Irenaeus fills his exposition with many passages from the books of the Old Testament prophets and uses biblical stories to accuse the unrepentant and the 'false preachers' in his age.

The most interesting section of this book is the six-page account of *Wunderzeichen* witnessed and recorded by Irenaeus himself, which occurred after midnight on 18 February 1564.[48] According to his account, some clouds gathered from north and east, and people saw fire burning in them. The clouds started radiating bright rays of light in the direction of the south and the west, and the fire and rays reminded Irenaeus of sparks flying in the workshop of a refinery. The sight was so dreadful that it looked like the Last Judgement was coming or God was punishing this world as had happened to Sodom and Gomorrah. Like other citizens, Irenaeus jumped from bed and rushed to the *Holzmarkt* to see what was going on. He told people, 'God has something in His mind so that he shows us this dreadful sign'. Irenaeus climbed up the tower of St Peter and Paul's with other church staff to see what would happen next. Soon a black cloud (in the shape of a long, wide block) appeared in the sky and remained still in the air behind the tower of St Andreas's Church in the Market Place. From the strange cloud, a group of knights armed with spears came out and started fighting each other. Blood was running under the fighting knights, and sparks and smoke 'like from cannon or guns' were also observed. But soon 'a gruesome fiery pool' appeared in the sky, and at the end the whole army fell into it. From the 'fiery pool', great smoke, steam and mist came out. It was a fearful sight to see, Irenaeus states. He lists the names of eight people who witnessed this *Wunderzeichen* from the church tower, including two *Rathsverwante*, the teacher, the chaplain and the sexton of St Peter and Paul's.[49]

To clarify what God wanted to convey to the people by showing such strange phenomena is the topic of the second part of the book. This part is divided into seventeen sections, and Irenaeus covers various topics, including God's anger regarding people's sins, the signs of his wrath, and the meaning of celestial wonders in general. The conclusion Irenaeus makes is a familiar one: these strange phenomena in the sky are not only the sign of divine punishment but also of the coming of the Last Judgement.

Irenaeus first reminds readers that God is a righteous judge who hates all sins, errors and transgressions (Section 1). The cause of human sins is Satan, 'a beginner and a founder of all sins'. After he was thrown out of heaven, day and night Satan plots how to tempt human beings and how to make them commit sins. But there are two additional factors helping Satan: our depraved nature due to the Fall, and the evil examples of the world (Section 2). God cannot tolerate to see human beings committing adultery with the devil and punishes sinners, while those who repent and admit their own sinfulness are forgiven and receive grace (Section 3).[50]

In the next section, 'God Announces Punishment for Sin through Law and Signs', Irenaeus presents his new knowledge of celestial wonders and explains the meaning of these phenomena (Section 4). First, he argues that whenever God decides to punish a land because of the people's error, idolatry and sins, God first warns the people by preaching his Word through his prophets. Moreover, God shows dreadful signs such as '*Chasmata*, fiery flames, armies, conflicts or battles and other dreadful visions in the sky. In addition, comets, terrible weather and floods on earth, terrible storms, etc'. These *Zornzeichen* warn people of the coming of punishment and serve as *Busspredigten*, like the signs urging Lot to leave the sinful city of Sodom.[51] Then Irenaeus lists various examples of God's warning signs in the Old Testament, starting with the warnings given to the Egyptians through Moses. The strange celestial signs that appeared before the destruction of Jerusalem, a familiar reference in *Wunderzeichen* literature, are also recounted in detail. Irenaeus next turns to some examples of celestial phenomena which appeared in his century that also foretold disasters. The first example he presents is the one he already mentions in the *Symbolvm Apostolicvm*: the sign that appeared in the sun when Elector Johann Friedrich was captured. Praising the steadfastness of the old Elector, Irenaeus interprets the sign in the sun as the spiritual darkness and blindness that covered the Lutheran faith following this event.[52] The list of *Wunderzeichen* cases continues, though they are not presented in chronological order: the signs that appeared before the Peasants' War; the storms in 1564, which were regarded as a portent of foreign invasion; the mysterious battle scene and other strange phenomena before the deaths of Elector Moritz of Saxony and Margrave Albrecht of Brandenburg-Kulmbach in 1553; the 'fiery balk or tree' ('fewriger Balcken oder Baum') in the sky before the invasion of the Turks in 1520; the comet and the vision of a hand holding a sword that foretold the siege of Vienna by the Turks in 1529; and the fiery sky in December 1561, followed by the Muscovite invasion. Many celestial signs also appeared in 1562 and 1563 before the outbreak of a war between Sweden and Denmark.[53] In addition to these *Zornzeichen*, another dreadful one appeared in February 1564, exactly eighteen years after Luther's death at Eisleben. For Irenaeus, this cannot be coincidental. 'Without doubt' the time of God's terrible punishment for sin in Germany is coming.[54]

How did people react to these 'sermons' of *Zornzeichen*? Alas, 'the world always remains the world' (Section 5). People remain in sin and let *Zornzeichen* make noise as if they were mere wind. They are like children: although they fear and tremble when they see a man with a rod, they soon forget it and return to the same behaviour.[55] They cannot stand to hear the true, righteous preaching, and they hate, ridicule and persecute true preachers (Section 6). Here Irenaeus quotes twenty-five Scripture passages (mainly from Isaiah, Jeremiah and Hosea) in which God accuses human sinfulness and expresses his wrath through the words of the prophets, adding a warning that these words are not only addressed to the Jews in the Bible but can also be applied to his age.[56]

What makes matters worse is that *Heuchelpropheten vnd Suppenprediger* (hypocritical prophets and 'soup-preachers' who are serving their own bellies) are encouraging unbelievers and instigating them to persecute true preachers (Section 8). While true prophets warn people, false prophets please them with vain hope, promising 'pax & securitas'.[57] God patiently waits for the people's conversion and continues to warn them, but as they show no sign of repentance, God finally punishes them, as the examples of Noah's Flood, Sodom and Gomorrah, the Babylonian captivity and the destruction of Jerusalem all show (Section 9). Germany will experience God's wrath through 'great pain, gruesome whipping and plague, from the Muscovites, the Turks, the pope and other folks'.[58] Once the time of punishment comes, any prayer will simply fall on deaf ears (Section 10).

The title Irenaeus chose for Section 14 is 'A Great Misfortune Will Come to Germany'. Though Germans have received the light of God's Word, he laments, they keep indulging in their sins.[59] The more time passes, the more the world becomes godless, as if all the devils in hell were liberated and now dominate and bind people.[60] The authorities are supposed to suppress these vices, but instead they overlook them and consequently encourage evildoers. Indeed, they are helping false preachers to attack the true and steadfast. Adiaphorists, Calvinists and Synergists are explicitly named as examples of false preachers who are receiving the authorities' support.[61] Now *Straffzeichen* (signs of punishment) appear in the sky, on water and on earth as a final warning, and it is clear that a dreadful punishment is at hand. Here Irenaeus quotes Luther's prophecies on the destruction of Germany to buttress his claim. Luther repeatedly predicted that terrible disasters would befall Germany soon because of the German people's ingratitude towards

God. When the Interim Crisis broke out, many Gnesio-Lutherans believed that the gloomy prophecy of the 'Third Elijah' was fulfilled, and they collected Luther's various prophetic remarks about the future of Germany from his writings.[62] For Irenaeus, Luther's prophecy is the most effective weapon to 'prove' the validity of his view. In these quotations listed by Irenaeus, Luther laments that Germans will suffer from 'terrible darkness and disasters'. Though no country other than Germany has listened as much to God's Word, people let it pass without gratitude and respect. Dear Germans, Luther cries, take God's grace while it is still there. God's Word can leave and will not return. Once God's Word was in Judea, but it is no longer there. Paul brought it to Greece, but now the land is dominated by Turks. Although Rome and other Latin lands once enjoyed it, now the pope is ruling them. 'And you Germany cannot think that you can hold it forever'.[63]

Where will the rod of God come from? Irenaeus was sure that it would come from the north (Section 15) because Ezekiel 38–9 warns that the army of Gog will come from the north to punish the Israelites. Irenaeus quotes Luther's exposition on Ezekiel, Genesis and Psalm 120, which states that the Gog can be the Turks, the Tatars and the Muscovites. Irenaeus is inclined to regard the Muscovites as Gog, and he adds an extensive comment on Luther's view concerning the danger of foreign invasion.[64] This concern is understandable because, as we read in Chapter 1, the news of the 1558 Muscovite invasion of Livonia was still fresh in people's minds.

It is no wonder that Irenaeus exclaims that the Day of the Last Judgement is at hand, as all the signs of the Judgement foretold by Christ himself have appeared (Section 16). The first sign fulfilled is that the Gospel was preached throughout the world because it is already preached in Asia, Africa and Europe. The second sign is that many sects, heretics, false teachers and *Schwermer* (enthusiasts) are appearing and spreading their errors. Calvinism especially is now poisoning Germany, and few theologians remain immune to its influence. The third sign is that all people are absorbed so much in their sins that one can hardly find faith, love and truth on earth. God is already punishing Germany with a number of rebellions and wars, inflation, pestilence and many kinds of sickness to urge people to repent. Moreover, all creatures in the natural world are showing signs of the end. They 'became tired and weary of serving the godless world as servants', and they are 'waiting, being anxious, and longing for salvation and revelation to God's children and will be renewed and serve only the elect'. Now the sun, the moon and the stars are eclipsed and show the signs of God's wrath. The earth trembles because it hates to carry so many people committing evil deeds. Water and wind roar and blow over people's *Sicherheit* and arrogance, as if to say, 'When it is permitted we will swallow you at once'. Thus all elements in nature are showing God's dreadful punishment and *Zornzeichen*. Though this section is based on Paul's words in Romans 8, which state that all creatures long for the time of salvation, Irenaeus presents a very fearful image: that of nature grumbling over the sins of humankind and longing for the moment to swallow it. Thus it is clear that the world is quickly approaching its end. Let us call God and pray that he will grant us the Holy Spirit and enlighten our hearts so that we can discern and acknowledge our sins, receive forgiveness and be led to true conversion (Section 17).[65]

Irenaeus's argument can be called a compendium of contemporary *Wunderzeichen* messages, and it does not show much difference from other authors' works; but we can perceive several distinct characteristics of Irenaeus's style. The first and foremost is that Irenaeus poses as a prophet of God's wrath, like the ones in the Old Testament. Most Scripture verses and biblical stories are quoted from the books of the Prophets such as Jeremiah and Isaiah, while only a few come from the New Testament. There is no Word of Christ, except his gloomy prophecy about the destruction of Jerusalem in Luke 19. This shows that this book was written as a *Gesetzpredigt*, which aimed to terrify readers by the Law and God's wrath so that they would experience spiritual despair and humbly seek God's mercy. Irenaeus's tone suggests his genuine frustration with 'epicureans', false preachers and the thoughtless authorities supporting them. His attacks on people's vice and the authorities' neglect of their duty sound conventional, but his repeated use of the phrase 'all devils go free' ('Alle Teufel sind fast ausgelassen') when he describes the sinful condition of the world reflects his gloomy view of human nature.

Compared to the *Symbolvm Apostolicvm*, Irenaeus's use of wonder stories has expanded in this work, although the number of cases he quoted is still limited and the description of *Wunderzeichen* listed in Section 4 is not long. After he published this book, Irenaeus must have grown aware that *Wunderzeichen* stories were effective tools of his ministry in attracting people's attention. Two years later, Irenaeus compiled a book that established a new genre of literature – a combination of religious polemic and wonder book.

Wasser-Spiegel (1566)

Ireaneus's *Wasser-Spiegel*, published by Andreas Petri of Eisleben in 1566, is an idiosyncratic book. If we summarize its contents in one sentence, it can be described as a motley collection of stories about God's providence manifested through water and floods, based on biblical tales, historical anecdotes, folk tales, contemporary news and the author's personal experiences. Although these stories are roughly classifiable into three categories, the order is quite arbitrary. Nonetheless, Irenaeus's goal is always clear: he intends to convince readers that God is inflicting watery punishments on unrepentant people, especially false teachers. Stories of wondrous events are narrated in order to lead readers to this realization.

Wasser-Spiegel consists of one introductory part followed by three long chapters: 'Zorn-spiegel' (Mirror of Wrath), 'Creutz-Speigel' (Mirror of the Cross) and 'Trost-Spiegel' (Mirror of Consolation). The introductory part consists of a dedication to Count Georg of Gleichen, Martin Luther's brief preface to Ambrose Moibano's commentary on Psalm 29, and Irenaeus's exposition on the meanings of *Zornzeichen*. The 'Mirror of Wrath' is a collection of stories of floods sent by God to punish people for their sins and to act as warnings of further impending disaster. The 'Mirror of the Cross' recounts various stories about floods that God had sent to punish false teachings, and here Irenaeus emphasizes the analogy between floods and the spread of false teachings. Finally, the 'Mirror of Consolation' catalogues narratives of God's grace as shown through water, including the miraculous rescues of flood victims.

Irenaeus fills his dedication to the Count with exactly the same arguments he wrote in the *Abdruck*: the world is filled with sin, and the authorities do not suppress these evils, while true preachers who denounced these vices are persecuted by false brothers. The more interesting section for us in the introductory part is Irenaeus's exposition on God's *Zornzeichen* (sign of wrath). Irenaeus first compares God with a good father: a true father punishes his children, but he first shows his rod to warn his children and press them to amend their ways. God threatens people first through the preaching of his Law and then through *Zornzeichen* like the father's rod. Irenaeus catalogues various types of signs of wrath in the following paragraphs, and his lengthy list shows that he had meanwhile extended his previous knowledge of wonders in the sky: 'multiple suns and moons', 'great eclipses and darkness', 'comets', 'rods', '*Chamata*', 'fiery flames', 'fiery light beams', 'blood rain', 'fiery sword', 'fiery crosses' and visions of 'armies and battle' appear. Sometimes 'cries of lamentation' are heard in the sky. Terrible storms, thunder and lightning, strong wind, the 'rain of stones', 'the rain of frogs' and floods can also be counted as *Zornzeichen*. Battles of birds, earthquakes, monstrous births and 'crosses and strange signs [which appear] on cloth' can also be added.[66]

After discussing God's wrath on human sin expressed through these signs, he turns to his favourite topic: the punishment of evil rulers, especially those who persecuted true believers. The story of King Nebuchadnezzar of Babylon is chosen as his first example, and Irenaeus compares the king's persecutions of the Israelites with the Interim. After a lengthy explanation of this analogy, he points out that Nebuchadnezzar's fall into madness was foretold by God in a dream. Warning signs that also appeared prior to the deaths of King Belshazzar and Julius Caesar are recounted, followed by the portents that foretold the destructions of Jerusalem and Constantinople.[67] This exposition of *Zornzeichen* can hardly be called well organized, but the flow of Irenaeus's argument reveals the movement of his thinking pattern. He first addresses wonders in the natural world, then God's punishment of people's sins, then his wrath against evil rulers and the Interim (false teachers), and finally returns to *Wunderzeichen*. Wonders, divine wrath, sin, persecution and false teachers are always intertwined in Irenaeus's mind, and throughout the following chapters, this same combination appears repeatedly.

The first topic of the 'Mirror of Wrath' is Noah's Flood, which supposed to have occurred in the 1,657th year after Creation. Seeing people's epicurean life as a cause of the Flood, Irenaeus uses the story to thunder about the lifestyle of the German people of his time. People are indulging in gluttony and heavy drinking everywhere in Saxony, Franconia, Thüringen, Meissen, Silesia, the Marck and Pommern. After describing the vice of these bad habits, Irenaeus coldly states that drowning in a flood in their houses is an 'appropriate punishment' for people dedicated to such an 'Epicurischen vnd Sardanapalischen leben'. But such a sinful lifestyle is not monopolized by the German people alone. People in the Netherlands, Holland, Brabant and Flanders are also known for their arrogance ('Vbermut') and luxury ('Pracht'), and that is why God often inflicts these lands with floods.[68]

In the following section, Irenaeus recounts various examples of storms, floods and other unusual incidents, and he points out that storms and floods struck Germany every year throughout the 1560s. He insists that when we see floods we should consider them as 'the mirror of wrath and punishment' shown by God and his calling for repentance. However, people forget God's rod as soon as the flood withdraws, and that is why floods have occured one after the other in recent days. Irenaeus's own experience of a violent storm in 1565, which may have inspired him to compose the *Wasser-Spiegel*, is described in detail in this section. A dreadful thunderstorm struck Eisleben during the night before Easter, and the inhabitants of the city were filled with such fear that they thought the Last Judgement might be coming. During the storm, Irenaeus recalled Luther's remarks concerning the Last Judgement, and he considered that since this was the night before Easter, maybe Christ would return to the earth and bring an end to the godless world in a 'Universal Easter'. The storm caused flooding, and many houses were immediately filled with water. The house of Bartel Voigt, one of Irenaeus's parishioners, was heavily damaged, and his baby would have died in the cradle if God had not granted mercy and saved it. Irenaeus detailed this miraculous rescue in a later chapter and expressed his sincere thanks, believing that one can learn a good spiritual lesson from this dramatic rescue: while the flood is the 'mirror of wrath' proclaiming God's punishment, it is also a 'mirror of consolation' for those who fear the Lord.[69]

The rest of the chapter is filled with stories of disasters, bloodshed and divine retribution. In several cases, floods were followed by other kinds of misfortunes such as plague, hunger and wars. For example, the terrible storms in Bavaria in 1529 were followed by the invasion of the Turks and the Siege of Vienna. This topic gives Irenaeus a good opportunity to terrify readers with the prospect that God would use the Turks, the Tatars and the Muscovites as instruments of his wrath. The possible connection between the prophecy of Gog's invasion in Ezekiel and these heathen people, a topic amply discussed in the *Abdruck*, is invoked again. In addition to Luther's gloomy prophetic remarks on this topic, Johan Hilten's prophecy that Europe would be subjugated by 'terrible, tyrannical people' in 1600 is quoted.[70] Thus the 'Mirror of Wrath' concludes with a depressing prediction about the future of Europe.

The second chapter is the highlight of the *Wasser-Spiegel*, because here Irenaeus freely expresses his passion as a confessional polemist and attacks the theological 'errors' prevailing in Germany, mainly the Synergists and the Calvinists. This part is named 'Mirror of the Cross' because it focuses on the interconnection between floods, false prophets and the persecution of true believers. Irenaeus connects these topics by emphasizing the analogy between floods and false teachings. Like floods, heresy and errors spread immediately, cover the land, and swallow people. Floods leave vermin, snakes, frogs and other poisonous animals on land after the water goes out, and these abominable creatures can be compared to the false teachings poisoning human souls.[71] Moreover, history tells us that floods, earthquakes or storms occur prior to the rise of a new *schwermerey* (enthusiasm or madness). Through these analogies, Irenaeus manages to connect the imagery of water, flood and erroneous faith.

After recounting the history of various theological errors, from Origen to the papacy, his argument turns to *schwermerey* among contemporary Protestants. His first target is the Sacramentarians in Switzerland, and he states that the reason why so many floods occur in Switzerland is that these floods are sent as punishments for the 'Sacramentarian poison' spread by Zwingli, Calvin and the other theologians. These 'poisonous hordes' are spreading not only in Germany but also in France, England, Poland and Italy. The Netherlands were well known as a place with many floods, and for Irenaeus the reason is clear: 'Satan flushed gruesome vermin and the poisonous hordes of Anabaptists, Sacramentarians and other *schwermer* into the place as God's righteous justice and punishment'.[72]

However, Irenaeus's primary concern is with the condition of Germany, and he refocuses his aim onto the Synergist theologians such as Viktorin Strigel and Johann Stössel. Irenaeus first reminds readers that there were many floods in Thüringen in recent days, and he uses this fact as an introduction to his argument against the Synergist ideas that are 'flooding many churches and schools'. Many pastors were misled by this 'water-snake or quacking frog', he laments. On the next eight pages, Irenaeus thunders against the teachings of the Synergists and compares these theologians with violent wolves. Quoting Fincel's book relating the story of a mysterious wolf appearing in Weimar in 1557, he explains that this wolf symbolizes a 'spiritual wolf', foretelling the rise of the Synergists in Saxony. Such a beast must be expelled by the 'clear light of God's Word' and 'barking dogs, that is Christian and passionate preachers' (in other words, like him).[73] Irenaeus's intense anger with the Synergists is understandable, as he had to leave his prestigious post at Weimar in 1562 due to his refusal to accept the theological view of Strigel.

After lengthy invectives against his old foes, Irenaeus invokes the memory of the plight of the pastors expelled by the Synergists and connects it with the story recorded by Fincel of a baby in Altenberg crying in the mother's womb. This gives him an opportunity to return to stories of wonder, and he quotes a number of *Wunderzeichen* cases from Fincel's book, including various signs in the sky (1525, 1532, 1554). For example, a blood-red cross was observed shortly before the death of the pious Duchess Sybilla of Saxony in 1554. After recounting the appearance of a spring of blood in the castle garden of Weimar in 1555, Irenaeus states that the meaning of these signs is the same as the warnings given to the Jews.[74] Many floods which occurred between 1561 and 1566 can be added to this list; surely God's punishment for people's scorn and the persecution of his Word is coming.

After his attack on Synergism, more discussions based on the analogy between false teachings and watery disasters follow. As he recounts several stories of rulers who died in the water, Irenaeus states that it is more terrible to see a ruler fall into theological error. 'Unfortunately', he says, 'in our age' one prince accepted Calvinism and flooded his whole territory with a 'horde' of Calvinist teachers. Obviously this prince is the Elector of Palatine, who leaned heavily towards Calvinism in the 1560s. As he describes the seductive power of Calvinism, Irenaeus suddenly quotes the legend of the Pied Piper of Hamelin, which he obviously learned from Fincel's volume. Following Fincel's interpretation, Irenaeus interprets the Pied Piper as a devil and laments that today false teachers are misleading youths with sweet teaching: 'But in our age Satan creeps in with his *Schwarmmerlern* and Calvinist seducers, Synergists and others, plays the pipe sweetly, captures lovely youth with sweet words and brilliant speech, surprises them, and makes them look pleasant with beautiful Lady Philosophia.'[75] This description suggests his conviction that the allure of Synergists and Calvinists is in their rational explanations of theology, agreeable to Lady Philosophia but not to Christ. Various historical examples of religious persecutions that broke out after the floods are presented, though Irenaeus does not discuss the floods themselves. What he wants to stress is that history shows that false believers always persecuted the true Church.

The last chapter of the book, 'Mirror of Consolation', is about God's grace as shown through water. This part is a collection of inspiring stories related to water, including allegories, devotional texts, legends, folklore and contemporary reports. In the concluding part of this chapter, Irenaeus states that the increase of floods in recent days is rather a consolation for true Christians because this is a sign that the Last Judgement is coming soon. The contents of the following paragraphs are largely identical to Section 16 of the *Abdruck*;[76] the weariness of creatures in nature to serve the sinful world is also discussed. It is clear that the end of the world is really at hand, but unlike the godless, the true believers, who take the warnings of God seriously, are consoled by the signs in the sky, on the earth and in the water. They raise their heads because the day of salvation is near, as Christ our Saviour himself promises in Luke 21.[77] With this assurance, Irenaeus closes his *Wasser-Spiegel*.

For modern readers, the structure of this unique work seems disorganized and confusing. A vast amount of stories and Irenaeus's own theological arguments are presented in loose connection with the imagery of water and floods, though they are sometimes not related to water at all. It is often difficult to follow Irenaeus's discussion, as he constantly moves from one topic to another. However, his thinking always returns to two points: God shows his wrath for people's vices (including erroneous teaching) through various phenomena in nature; and because false preachers are misleading people in this world, true believers are always persecuted. These two points, already present in the introductory part of the book, are at the heart of Irenaeus's thought.

Irenaeus's choice of floods shows that he grasped the common people's daily concerns and exploited them for his theological agenda. His impression that storms and floods are increasing in his days was not mere rhetoric but was shared by many contemporaries. In Chapter 1 we saw that the climatic anomaly brought on by general cooling caused many floods in the latter half of the sixteenth century. The statistical data presented in flood research by Rudolf Brazdil, Christian Pfister and other historians show a marked increase of floods (often the worst of the century) along the Main, the Pegnitz and the Elbe during the 1560s. In 1566 the flood of the Rhine marked the highest overflow in the Basel region in the last 500 years.[78] Eisleben also experienced flooding (though not caused by an overflowing river), leaving a strong impression on Irenaeus's mind. People's shock and bewilderment at such a sudden increase of unusual weather is well described in Spangenberg's *Mansfeld Chronicle*. Considering these facts, Irenaeus made a clever choice of topic to draw people's attention.

Conclusion

Irenaeus's three books discussed in this chapter show the development of his interest in wondrous phenomena, as well as his ability to use these in support of his polemical agenda. The insertion of several *Wunderzeichen* cases in his argument against astrology in the *Symbolvm Apostolicvm* shows that Irenaeus was interested in celestial wonders from the beginning of his clerical career. However, in this book he had not yet directly connected stories of wonder with his theological agenda.

The *Abdruck* was his first work that used *Wunderzeichen* stories to support his theological messages. The list of strange celestial phenomena listed in this book suggests that Irenaeus developed his knowledge and interest in *Wunderzeichen* after he published the *Symbolvm Apostolicvm*. However, he still used them in the same way other preachers did in their 'repent or disaster' sermons. His attack on people's scorn for God's Word is rather conventional, and his warning of the apocalyptic invasion of Gog – especially the Muscovites – was hardly original.

As we saw in Chapter 1, a number of preachers, including Andreas Musculus, adopted this warning of foreign invasion in their *Wunderzeichen* messages. Indeed, like Musculus, Irenaeus got this idea from Luther's prophecies about the future of Germany; it is no wonder that he connected them with the invasion of the Muscovites, the major news of the day. In the *Abdruck*, Irenaeus poses as a Jeremiah-like prophet and attacks people's immoral lives and contempt for God's Word. The fact that he quotes mostly from the Old Testament shows that he intended his book as a *Gesetzpredigt*. Although the complete absence of Christ's Gospel from the book gives us a gloomy impression, here Irenaeus is following the formula of Lutheran pedagogy of the conversion process. Nonetheless, his anger against false preachers and authorities persecuting true believers, which he repeatedly invokes, is unusually intense. It is easy to imagine that Irenaeus's bitter experiences of losing his post at the court instilled in him a pessimistic attitude towards secular authorities' interference in religious matters.

It seems that the *Abdruck* was well received by readers, as Irenaeus compiled the *Wasser-Spiegel* two years later and filled the book with numerous eye-catching and sensational tales. Obviously he had learned that stories of wonder could draw people's attention, and he decided to exploit this appealing power for his polemical purposes. In this new book Irenaeus combined two literary forms – religious polemic and wonder book – and created an original genre. The imagery of water and floods was used to attack his theological opponents, mainly Synergists and Calvinists, as false teachers who were inviting God's watery punishment. At the same time, Irenaeus inserted a great number of interesting anecdotes to maintain his readers' interest.

Though their contents are largely different, both the *Abdruck* and the *Wasser-Spiegel* share the same message, consisting of the following three points: God is threatening the world with his wrath, and if people do not amend their way of life and discard erroneous teachings, terrible punishment (most likely foreign invasions) will befall Germany; most people in this world are following false prophets; and the true servants of Christ are persecuted. Surely these false teachers will invite God's retribution, and various signs (including *Wunderzeichen* in nature) show that the end of the world is at hand. To warn people of these dangers was Irenaeus's lifelong mission, and this conviction was already expressed in his introduction to the *Symbolvm Apostolicvm*.

Despite his love of disputations, as a pastor of the historical Eisleben church, it seems that Irenaeus was respected and loved by his parishioners. He was recalled to Weimar as a court preacher and left his pastoral post in 1566. When Irenaeus's successor died the next year, the congregation of St Peter and Paul's wanted its former pastor back, but Duke Johann Wilhelm did not allow his return.[79] This suggests that Irenaeus was popular among his parishioners and also cherished by the duke, despite his sceptical attitude towards secular authorities. The three books we have examined offer several clues to explain his popularity. There is no doubt that he was a man of strong curiosity; he probably loved to tell people the stories he collected, and his rich imagination and use of creative words allow us to imagine that Irenaeus was a good storyteller. His adoption of flood motifs suggests that Irenaeus also had a keen sense of his parishioners' concerns. His endless curse of false preachers may be distasteful to us, but since aggressive disputation was common among sixteenth-century theologians, his parishioners may have simply perceived it as an expression of their pastor's genuine passion for purity of doctrine. Irenaeus's ousting from the ducal court showed them that their pastor was a man of principle, while his recovery from dangerous pestilence and his recalling to Weimar after the dramatic fall of the 'persecutors' (some of whom were executed as traitors as the result of the Gotha Rebellion) might have been seen as providential favour. In addition, Irenaeus's lengthy reference to the baby rescued from the Eisleben flood shows that he was a conscientious pastor with warm concern for his parishioners. In 1564 he published his sermon on the story of Martha and Mary in Luke 10, originally delivered at the burial of an Eisleben widow. In the dedication of the printed text, Irenaeus affectionately recalls his own mother, who died a month prior at eighty years of age, and states that his mother was both Mary, who did not neglect God's Word, and Martha, who took care of her late husband and seventeen children.[80] This warm statement also shows that he was not merely a 'rabid' theologian but also a man with an interesting life. However, within several years his promising career came to an abrupt end, and the term *Exul Christi* (exile of Christ) became his identity. The cause of this change is original sin, the topic that he emphasized in his anti-astrology arguments in the *Symbolvm Apostolicvm*.

5 IRENAEUS AGAINST THE CONCORD: POLEMICAL USE OF *WUNDERZEICHEN*, II

In this chapter, we examine Irenaeus's two wonder books and two additional related publications he wrote in the 1570s and 1580s, and we analyse how he used stories of wonder to promote the Flacian view on original sin and his own attack on Andreae and the Formula of Concord. The first wonder book, *Prognosticon Aus Gottes Wort*, which was inspired by the great comet of 1577, combines a lengthy catalogue of celestial *Wunderzeichen* and other kinds of prodigies with his refutation of the Concord. The second work, an exposition of Christ's prophecy concerning signs of the End Times, is also a spiteful polemic against Andreae, but here he shows an interest in both celestial wonders and monstrous births. The third text does not refer to wonders themselves, but it indicates the direction in which Irenaeus's interest in monsters is going, in accordance with his polemical intention. In the last book, *De Monstris*, Irenaeus mobilizes all the skills of creative writing he developed during the past two decades and masterfully transforms stories of deformed children into emblems of the total depravity of human nature and the inconsistency of contemporary theologians. Before looking into these works, it is necessary to grasp what was at stake in the Flacian Controversy. Irenaeus's polemical activity in Mansfeld will be examined, because it was there that he forged his identity as an *Exul Christi* and his sense of being a 'remnant' of the true believers, which he retained throughout the rest of his life in exile. Only then can we comprehend why Irenaeus was so eager to draw readers' attention to *Wunderzeichen* stories and use them as an introduction to his polemics. For him, the struggle was not a mere rivalry between theologians. It was a cosmic battle between Christ and Belial, and *Wunderzeichen* show God's will with regard to false teachings prevailing in Germany.

The Flacian Controversy and Irenaeus's Polemical Activity in Mansfeld

The Flacian Controversy started with Flacius's criticism of the Synergist position of Viktorin Strigel, a fellow professor at Jena University. The difference between the theological view of Strigel and that of Flacius became manifest during a public disputation held at Weimar in 1562. At the disputation, Strigel insisted that despite the Fall, the human will still retained the capacity to collaborate with God's work by assenting to God's gift of grace. Through original sin, humankind lost the power to be good, but this does not mean that the original substance (fundamental identity) of human nature itself has changed. Employing Aristotelian terminology, Strigel concluded that original sin must be an 'accidental' element of human nature.[1] Flacius instead insisted that Scripture clearly states that the Fall brought about a substantial transformation of human nature, not only an accidental change. Through the Fall, 'the image of God, has not only vanished, but now is changed into its contrary, indeed the image of the devil'.[2] Thus Flacius suggested that original sin became the very substance of human nature, and this meant that there was no room for the human will to collaborate with God's work of grace. At first, many Gnesio-Lutheran theologians were favourable towards Flacius's position. However, after Flacius articulated his view and published it in 1567, Gnesio-Lutheran theologians began to notice that the Flacian position contained several logical difficulties, and they soon split into two groups. Flacius tried to defend his position by claiming that the substance could be divided in a higher and a lower form, and the latter could still retain the original perfection despite the former's corruption.[3] However, the majority turned against the Flacian understanding, while at the same time they rejected Strigel's Synergist stance.

Why was this argument over original sin so significant to Lutheran theologians? Since the Interim Crisis, Gnesio-Lutherans had recognized that most 'false teachings' such as the Leipzig Interims and the Synergist view had originated from limitations placed on Luther's principle of *sola gratia*. If human nature was not completely corrupted, then it was possible to argue that human will could still contribute to one's salvation in some way, and this left the door open to theological error. In the words of the pastors of Mansfeld, it

> will open a huge window through which the Papists would adorn and defend their teachings on natural integrity ... the Synergists their means of preparation for grace, that is, passive acceptance; and the Maiorists their necessity of good works for salvation.[4]

The Flacian understanding could effectively close this window, and this was the great attraction of the 'substance' interpretation of original sin.

Flacius and his followers insisted that their view was firmly based on Luther's writings. There is no doubt that Luther emphasized the extreme seriousness of the Fall. However, as Christman discusses, Luther considered original sin a 'damaged relationship' between God and man.[5] In many cases Luther loved to discuss his theology without using the terminology of philosophy, and this attitude often posed difficulties for his followers as they tried to forge Luther's faith into solid doctrine. As Christman states, '[b]ecause the language of broke relationships was not the language of academy, already prior to Luther's death the tendency in Protestant circles had been to "translate" his theology back into an academic conceptual framework'.[6] Flacius's explanation was one of these attempts to 'translate' the dynamism of Luther's theology into philosophical argument, and the controversy he caused was one of these struggles to define which group was the true heir to Luther's theological legacy.

Soon the Mansfeld region became the centre of the Flacian Controversy over original sin. At first the pastors of Mansfeld were united in supporting the Flacian position, but by around 1570 a split rapidly developed. While many pastors in Eisleben, headed by the Superintendent Hieronymus Mencel, took the anti-Flacian position, the pastors of Tal Mansfeld, under the leadership of Cyriakus Spangenberg, supported the 'substance' view (ironically, during the emotional confrontations, Flacius's elaborate distinctions within human nature were forgotten). A colloquium of the two parties held at Eisleben in July 1572 failed to reach any consensus, and the split became decisive. Over the next two years the two parties engaged in fierce theological polemics.[7] In this heated atmosphere, Irenaeus arrived at Mansfeld and immediately participated in Spangenberg's campaign. He attended the ill-fated Eisleben Colloquium and explained his position to other pastors of Mansfeld.[8] In July 1573 Irenaeus was given a chance to preach to his former congregation at St Peter and Paul's Church in Eisleben.[9] He chose as his topic the spiritual wedding between Christ and his bride, the Church. The main message is that although human nature is utterly corrupt and damned by the Fall, Christ the groom mercifully gives her a wedding dress – his righteousness – to cover her miserable body.[10] In the printed version of the sermon, Irenaeus spends fifty-eight pages explaining the total depravity of human nature after the Fall, showing his desire to implant the Flacian position in his hearers' hearts.

Beside this sermon, Irenaeus published several pro-Flacian works during his stay in Mansfeld. In one of these works, titled *Apostasia* (1573), he describes various examples of apostasy from true teachings in history and warns readers of divine punishment. Irenaeus argues that when God grants grace to a country, he sends righteous teachers to reform the Church. However, after such true preachers leave the land or die, a general fall from true faith follows.[11] That is what happened to the Israelites after the death of Moses, and to Germany in the Interim Crisis following Luther's death. People became tired with God's

Word, sought new teachings (including *Separabile Accidens*), and persecuted true preachers. The condition of Germany after Luther's death is 'the last and the most dangerous fall from Truth'.[12]

The message of the *Apostatia* clearly shows Irenaeus's self-identification with the 'remnant' of true believers living in the midst of general apostasy. Robert Kolb points out that Gnesio-Lutherans developed a 'remnant' ecclesiology and took a defensive position, believing that 'the martyrdom of physical suffering and death may well come to those who give public witness to their death'.[13] Since the majority of Gnesio-Lutheran theologians took the anti-Flacian position, Irenaeus's sense of being a 'remnant' and his self-identification as a martyr had to be intense. The fact that he called himself *Exul Christi* after he lost his post in 1572 shows that he made martyrdom his identity. During his stay in Mansfeld, he added the title *Exul Christi* to his name in his publications.[14]

Unfortunately, the campaign by Irenaeus and other comrades came to an end probably earlier than they had expected. On New Year's Eve 1574 Tal Mansfeld was raided by the princes' soldiers, and Irenaeus had to escape on dark winter roads to avoid arrest. Together with Spangenberg and other pastors, the *Exul Christi* had to go into exile again.

Prognosticon Aus Gottes Wort (1578): Exposition of *Wunderzeichen*

We do not know where Irenaeus was residing in 1577, but he must have seen the bright comet that appeared in the sky in November. That year Andreae's Formula was finally completed and recognized as the confessional norm by the Elector of Saxony. For Irenaeus these two events could not be separated, because a comet and the success of false prophets were two clear signs of the end. Based on this conviction, Irenaeus took up the pen for his third *Wunderzeichen* book.

The full title of this work is *PROGNOSTICON from God's Word, Necessary Reminder and Christian Repentance Sermon Addressed to this Final Evil Time, to High and Low Classes in German Lands: On the Comet which was Seen from Martin's Day of the Year 1577, to the Beginning of the Year 1578. With Stories of Many Comets and Other Dreadful Signs and All Things which Followed Later.*[15] The printer's name is not known, but since the exiled Flacian theologians were maintaining support networks to sustain their faith, most likely Irenaeus found a printer through such networks. It seems that the sight of the comet reminded Irenaeus of the fiery sign he had seen in Eisleben thirteen years earlier, because the *Prognosticon* is virtually an amplified version of the *Abdruck*. The basic structure of the text and some of the chapters are a direct copy of the 1564 work. However, Irenaeus added two lengthy catalogues of *Wunderzeichen* cases. One is of comets that appeared from ancient times to 1572, and the other is of strange phenomena observed in the sixteenth century. Extensive theological discussions regarding the nature of original sin are also added to refute Andreae's claims.

The title *Prognosticon Aus Gottes Wort* shows Irenaeus's attempt to capture people's interest. *Prognosticon* meant a pamphlet of astrological predictions, an extremely popular literary genre. Irenaeus certainly expected that the use of the word 'Prognosticon' on the title page would easily catch people's eye. The term 'from God's Word' suggested that it was not a work of astrological literature but a book of religious inspiration, yet the phrase 'With Stories of Many Comets and Other Dreadful Signs and All Things which Followed Later' had to stimulate people's curiosity. Almost half the book offers theological polemics about original sin, but the title does not indicate this. It is likely that Irenaeus deliberately hid his intention from the title in order to secure more readers and avoid negative attention from the authorities.[16]

The *Prognosticon* consists of four parts: a description of the comet and other strange phenomena of 1577; Irenaeus's exposition on the meaning of comets and other *Wunderzeichen*, which is a conventional 'repent-or-disaster' sermon; a catalogue of historical examples of comets observed since ancient times; and 'Christian, necessary, and God's-Word-based admonition and repentance sermons addressed to this final age, for all classes in German Lands', which is a recycling of the main text of the *Abdruck*. Irenaeus opens his text with a brief report of various unusual incidents that occurred in 1577. In addition to the great comet on St Martin's Day, the catalogue includes two eclipses of the moon observed in April and September; great storms that struck Saxony in July and the casualties caused by the thunder; an earthquake; a spring of blood sprung from the earth; terrible winds and floods; dark mist covering the land; and a mysterious voice saying 'woe, woe, woe' heard in several places. In addition, an army of soldiers appeared in the sky, while two suns fought each other. Throughout history there has been no other age when so many strange phenomena appeared one after the other.[17] Surely God must have had a reason to show so many strange signs to people.

Irenaeus's interpretation of these phenomena, which he explains in the next section, is a conventional one. He explains that comets are created by God as signs of his wrath and punishment. Together with God's Word, these signs foretell future misfortunes that will befall people for their sins. The books of the Prophets (plus Luther's commentary) contain several passages suggesting that God shows strange signs to people before he inflicts punishments. History also shows that great misfortunes follow comets, such as the 'death of rulers, changes in religion and rulership ... war, bloodshed, the destruction of cities, the devastation of lands and people, pestilence and mass death, hunger and inflation, and punishments and plagues'. Even the ancient pagans regarded comets a sign of dreadful things. Thus God shows signs not for 'theatrical play or entertainment' but as a warning for us, as *Bussprediger* that urge us to repent, together with God's Word. Moreover, these signs were the harbinger of the Last Judgement.[18] Thus the comet of 1577 is 'a rod and a whip' placed before our eyes, and we must not disregard it but respond with repentance and improvement of our lives. Basically this message does not much differ from what he wrote in his previous *Wunderzeichen* writings.

This chapter is followed by a thirty-six-page catalogue of comets that appeared from Xerxes's time to 1572. This section describes sixty-eight cases of comets (fifty-five cases from the time before 1500, and thirteen from the sixteenth century). To each case, Irenaeus adds an explanation of what kind of disaster followed the appearance of the comet. One example is Halley's Comet of 1066, which appeared before the conquest of England by William of Normandy, and well known by its image in the Bayeux Tapestry. When the comet of 1312 appeared, the Templars were 'terminated in one day', and shortly after Germany suffered from severe epidemics and 'a third of the population died' (which may signify the Black Death in 1349). Concerning cases in his century, Irenaeus mentions the comets of 1500, 1506, 1516, 1523, 1527, 1531, 1532, 1533, 1535, 1538, 1539, 1541, 1554, 1556, 1558 and 1572. After a comet appeared in the summer of 1531 and remained in the sky for three weeks, war broke out in Switzerland and brought the death of Zwingli. The 1533 comet was followed by the Anabaptist Kingdom of Münster.[19] For the nova of 1572, Irenaeus uses the term *Cometischer Stern* and points out that wars occurred in France and Brabant, inflation and pestilence also broke out, and some princes in Germany died.[20] In the conclusion to this catalogue, Irenaeus solemnly admonishes readers that what will come after the 1577 comet is suggested by the condition of the Netherlands: 'Braband, unfortunately, sensed the meaning of the comet by being attacked and terrified by the Spaniards'. Because of our sinfulness, he warns, 'it is likely that Germany will not be free from approaching war, plague and all kinds of sickness and epidemics'.[21]

The sheer number of comet stories included here indicates that Irenaeus had considerably expanded his knowledge on *Wunderzeichen* since he completed the *Wasser-Spiegel*. After the catalogue, Irenaeus moved on to the fourth part of his book. This lengthy part is mostly identical with the main text of the *Abdruck*, though he altered the structure: while the *Abdruck* consisted of sixteen sections, the *Prognosticon* is divided into twenty-two sections.

A section completely changed from the *Abdruck* is the *Prognosticon*'s Chapter 6, 'God Forewarns his Punishment for Sins through Law and Signs' (this was Chapter 4 in the earlier work). This long chapter can be divided into two parts: the first consists of Irenaeus's discourse on God's warnings of punishment based on stories in the Old Testament, while the second catalogues various historical examples of divine warning. Though the discussion in the first part is largely identical to the *Ausdruck*'s Chapter 4, in the *Prognosticon* Irenaeus adds examples of prophets' calls for repentance. After these discussions, Irenaeus recounts the much-repeated story of divine warning: the destruction of Jerusalem.[22] This narrative serves as an introduction to his lengthy (forty-one-page) catalogue of *Wunderzeichen* cases. In Chapter 4 of the *Abdruck*, besides the Jerusalem story, only ten further examples of divine warnings are recounted; this is expanded to about fifty stories in the *Prognosticon*. Unlike the previous section, this catalogue includes all kinds of strange signs, such as unusual celestial apparitions, monstrous births, storms, battles of animals, mysterious voices and natural disasters.

Irenaeus starts his second catalogue with examples from the Middle Ages and points out that the rise of the two arch-enemies of true Christian faith, the papacy and the Turks, were preceded by strange signs.[23] However, most of the stories of wonder listed in this section are from his own century, mostly taken from Fincel's *Wunderzeichen*. For example, the story of the mysterious blood-red cross above the castle of Weimar, witnessed by Elector Friedrich and his son in 1517, appears at the start of Fincel's catalogue.[24] Irenaeus was obviously interested in this story, because he believed that the 'blood-red' cross could be regarded as a sign foretelling future persecution and the risk that true believers had to bear. Fincel simply stated that the sight foretold the religious split and 'danger to the House of Sachsen',[25] but Irenaeus explicitly stated that it signified not only the restoration of Truth by Luther but also the cross and persecution, which three princes of Saxony had to undergo.[26] For Irenaeus, the Reformation and the persecution of true believers were inseparable.

Irenaeus does not simply quote the events but occasionally adds his own comments on individual cases from Fincel's book. The story of the vision observed in the sky over Rome in November 1547 is another example. A blood-red cross, a rod and a black eagle appeared and remained in the sky for three days. Irenaeus regards the cross and the eagle as signs of persecution through the emperor's Interim and the rod as the persecutors and the supporters of the Interims.[27] In 1548 people in Jena saw a black large ball ('Kugel') appear and cover the sun like a cloth. According to Irenaeus, this was a sign that the Interim, signified by the black ball, would darken the sun, the symbol of Christ and his teaching.[28] Since Fincel wrote the same thing in his book, this interpretation is not originally Irenaeus's.[29]

However, Irenaeus applied the same view to other *Wunderzeichen* cases that Fincel did not comment upon. In June 1554 a similar phenomenon was observed in Jena. After the sun suddenly turned blood red, numerous small black balls appeared and went around the sun. To Irenaeus this signified change not only in the secular order but also in religious or ecclesiastical *Regiment*. The small dark balls signified false teachings (Strigel's theology), and in this case the sun symbolized the Christian teaching of original sin.[30] He then mentions the case of a landslide in Berka, Thüringen, which created some deep cracks ('Klaff') in the earth in 1555. From these cracks, bad, poisonous smoke and vapour came up, and at night many spirits ('Gespenst') were seen near the hole. In addition, a thunderbolt struck the earth near the city gate of Saalfeld on Easter 1555, creating a deep hole from which came the odour of sulphur. These cases seem to be pure natural phenomena, but Irenaeus read symbolical meanings into them: the poisonous smoke, vapour and stench signify that false teachings such as Strigel's *Declaration*, Johann Stössel's *Superdeclaration* and the *Accidens* would come.[31] Here he employs the scatological rhetoric commonly used by sixteenth-century polemics to describe the hideous image of false teaching.

After narrating these stories, he turns to the case of a wolf appearing in the city of Weimar in 1557, which he discussed extensively in the *Wasser-Spiegel*. As he wrote in his previous work, he states in the *Prognosticon* that the wolf represented a 'spiritual wolf'. This time, however, it means not only the Synergists but also the *Accidentz schwermer* who 'concocted ... the difference between the corrupt nature of humankind and original sin'. The case of the Eisleben *Wunderzeichen* in February 1564 is also recounted, but new elements are added to his narrative. Only three days after the appearance of this fiery sign, Archbishop Sigismund von Magdeburg held a lavish banquet at Mansfeld Castle. He drunkenly cut his beard for fun, and according to Irenaeus, the beardless Sigismund later issued a decree that the people of Magdeburg and Halberstadt had to cut their beards. Irenaeus laments that 'such frivolity had been unheard of'. For him, this incident was nothing but an expression of thoughtless contempt for God's *Zornzeichen*. Here Irenaeus overtly expands his imagination and states that faces with moustaches 'look like the Turks, the Tatars, and the Muscovites', and worries that this will become an overture ('Vorspiel') of foreign invasion by these peoples.[32] However, far worse than these invasions was the religious persecution in this land by the *schand Götzen Baal Accidens*. Irenaeus laments that the unity of teachers and believers in Mansfeld was destroyed by these 'shameful, seducing *Accidens*', and the misfortune and sad condition of the land was signified by many other signs.[33] Mansfeld was struck by a terrible hailstorm in July 1571, and people were so terrified that some of them thought that the Day of Judgement had come.[34] Irenaeus closes this chapter with a list of various phenomena of 1577, almost identical with the list he printed at the start of the text.

Prognosticon Aus Gottes Wort: Polemic against Andreae

Irenaeus's comments, which he adds to the *Wunderzeichen* stories in Chapter 6, clearly show his intention to attack the Synergists and *Accidens*; but he reserves his main attack of the *Accidens*, especially Andreae and the Concord, to Chapter 10, in the section about 'hypocritical and soup preachers' (this was Chapter 8 in the *Abdruck*). Irenaeus opens this chapter with a quotation from Jeremiah concerning false, hypocritical prophets who promised peace and good times even though people indulged in sins, while righteous and faithful *Bussprediger* threatened them with warnings of divine wrath. Unfortunately, this can also be said about contemporary Germany. In this world there is always conflict between Christ and Satan, with soldiers on each side. These parties cannot be reconciled, as the Bible and our daily experiences tell. After the Fall, God pronounced a sentence against the devil to the effect that 'I will put enmity between you and the woman, and between your seed and her seed' (Genesis 3:15). No one, 'yea indeed the Son of God himself', can overturn this sentence. Irenaeus castigates Andreae, who diligently worked for years to reconcile the two opposing parties: 'namely, the seed of the Woman and the snake; Christ and Belial; light and darkness; Truth and lie'. Irenaeus urges readers to ponder on Christ's own words: 'Do not think that I came to bring peace on the earth; I did not come to bring peace, but a sword' (Matthew 10). Since there must be *Rotten* and *Secten* in this world so that the righteous are revealed, conflict between the two parties will continue until the Day of Judgement. Christ also says that the way to heaven is narrow and only a few people can go on the path, while the way to damnation is broad. Irenaeus ridicules Andreae for teaching the contrary: the way to heaven is wide and many people will go.[35]

After castigating the idea of Concord, Irenaeus sets his target on the doctrine of original sin, the first article of the Formula. He states that this very first article is wrong and wonders whether we can expect any good to be built on such a vain and rotten foundation.[36] Throughout the following thirty-six pages, Irenaeus argues the twelve points which prove that the Formula's anti-Flacian interpretation of original sin is against the Bible, the catechism and Luther's writings.[37] Genesis clearly states that Adam was created in God's image; however, through the Fall he lost his holy and righteous nature and became a *Gegenbild* (opposite image) of the original image, in other words the image of the devil.[38] Genesis 5 says that Adam became the father of Seth 'in his own likeness', and Irenaeus points out that the Scripture does not say 'in God's image'. Certainly Seth was in the image of Adam after the Fall, and this means that Seth was not only sinful in accident but has also received the whole corrupt nature from his parents.[39] In Psalm 51 David laments that he is totally sinful even from conception,[40] and Irenaeus quotes Luther's comment on this penitential verse: sin is 'my whole nature, my beginning, and my conception, besides my word, work, thought, and later life. By nature I am a bad tree and a child of wrath'.[41]

Through baptism we are born again, become a new creature, and start a new life. Since it is 'new birth', this suggests that we are spiritually dead before baptism. Old Adam must die first because the former is original sin itself.[42] Therefore, Irenaeus argues, we believe that the Holy Spirit helps old Adam to be born again, gives him a new life, and enlightens him. For Irenaeus, to treat original sin as an accident is an insult against this dynamic life-giving act of the Holy Spirit that works through baptism.[43] These arguments, being only a part of Irenaeus's passionate discussion, clearly show that he regarded the theology of *Accidens* as heinous blasphemy that undervalued the seriousness of original sin. To divide human nature and original sin is nothing but (in Irenaeus's words) 'new alchemy', because our whole nature is against God and it is no longer possible to separate nature from original sin. In Irenaeus's eyes, the complete depravity of human nature is the very foundation of Christian faith; one has to depend on Christ only for salvation. The contents of the remaining chapters of the *Prognosticon*, which warn readers of divine punishment and the imminent coming of the End Times, are largely identical with the *Abdruck*.

The *Prognosticon* shows the development of Irenaeus's skill in the use of *Wunderzeichen* for his theological agenda. In the second catalogue of *Wunderzeichen*, he explicitly attempts to connect some of the *Wunderzeichen* with his polemical agenda. The location of each wondrous incident was important for the interpretation of the signs: Weimar, where a strange wolf appeared, became the stage of the fateful disputation between Flacius and Strigel, while Jena was the stronghold of Gnesio-Lutherans who later rejected the Flacian stance. Thus both places gave reason for signs of false teachers to appear. The fiery sign that appeared in Eisleben in 1564 was interpreted as a forewarning of the controversy and persecution in the Mansfeld region. For Irenaeus, these theological issues had cosmological significance, which is why God showed his concern about these issues through celestial signs. The connection between the first part of the book (wonders) and the latter part (religious polemics) is rather loose, and it is possible to imagine that some of the readers enjoyed only the first part and disregarded the latter. However, by combining religious polemics and *Wunderzeichen* stories, Irenaeus could expect many people to obtain the book, and the chances that they would read his arguments against Andreae and the *Accidens* would be greater than by publishing the same argument as a purely polemical work.

Erklerung des Euangelij Luce XXI Am andern Sontag des Aduents (1581) and *Contrafet Vnd Spiegel des Menschen* (1582)

In 1581 Irenaeus published *Exposition on the Gospel Luke 21 for the Second Sunday of Advent*. As we saw in Chapter 2, the topic of the pericopic sermon on the second Sunday of Advent was Christ's prophecy concerning the signs of the Last Judgement in Luke 21. However, Irenaeus had no intention of writing a sombre commentary on the Gospel for personal devotion. Though he was following the same basic format as other theologians' postils, his main aim again was to denounce Andreae, the Concord and its doctrine of original sin. In this book, *Wunderzeichen* fulfil only a minor role, but a new element appears in reference to wonders. Irenaeus opens his exposition with a discussion of what kinds of signs are supposed to appear prior to the end, but he quickly moves the discussion to the rise of false teachers and launches severe attacks on the Concord and his arch-enemy, Andreae. The fact that thousands of teachers and believers subscribed to Andreae's article on the *Schwarmreiche Accidens* is a 'manifest sign of the Day of Judgement'. After lengthy diatribes against Andreae, Irenaeus returns to the issue of apocalyptic signs. There will be wars prior to the coming of the end, and the present world is filled with wars (he was probably referring to the conflicts in France and the Netherlands). Here he abruptly states that 'therefore, comets or tail-stars and fiery rods, with red crosses threaten [us] and Dr. Jacob [Andreae] with his Formula of Discord ['Formel Discordia'] may not be a small cause of this'.[44] This has to refer to the comets that appeared in 1577 and 1580. The flow of his arguments seems to be confused, but Irenaeus's thinking is clear: apocalyptic signs include celestial wonders, and the comets recently observed were related with the arch false teacher Andreae.

This implication becomes clear in the passage immediately after the comet reference. Irenaeus quotes a Latin epigram about the double-headed baby born in Worms in 1578. This monstrous birth is interpreted as a sign of religious division ('Hoc signat sectum relligione gregem') and the conflicts between Jacob and Esau or Christ and Belial, accompanied by hostility and persecutions.[45] Irenaeus's reference to the Worms baby is noteworthy because the topic of his next wonder book is monstrous births, and it shows that he was already interested at this point in the parallel between the deformity of babies and the condition of the Church.[46] There are few new elements in the *Erklerung* because most of the arguments in this book are recycled from Irenaeus's earlier works. Interestingly, he refrains from displaying his extensive knowledge of *Wunderzeichen* here, probably because he thought that it would distract readers' minds. He does mention comets that appeared recently, but his reference to the Worms baby in connection to the Formula of Concord suggests that his interest in wonder is moving in a new direction.

This is attested by the fact that in the work published next year, *Image and Mirror of Mankind*, Irenaeus presents a parallel between fallen human nature and monsters. The goal of this work is to explain his view on the utter depravity of human nature, and here Irenaeus states that without Christ, human beings are 'hateful abomination in front of God', like an 'ugly, terrible monster or misbirths in front of our eyes'.[47] This passage is immediately followed by a familiar denunciation of '*Concordisten* and *Accidensschwermer*, new alchemists, distiller, divider and distinguisher of all kinds of excuse and evading speech, taken from twisted reason and all stinking holes of philosophers and sophists', which shows his rich vocabulary. In a later part Irenaeus presents a list of the various effects of the Fall, which include 'terrible *Monstra*, deformed misbirth ['Ungestalte Missgeburt']' and babies' cries of lamentation in their mothers' wombs – both phenomena were regarded as wonders.[48] What draws our attention is Irenaeus's growing fascination with misbirths, especially the analogy of corrupt human nature and monsters. It is clear that he sensed that this comparison would be effective in explaining his understanding of original sin. Two years later he published a massive volume based on this analogy.

De Monstris (1584): The Preface

In his previous works, Irenaeus covered three major types of *Wunderzeichen*: *Fewerzeichen*, comets and natural disasters (floods and storms). The only remaining type of major prodigy is monsters and monstrous births, an immensely popular topic in early modern broadsheets and tracts. Irenaeus did not overlook the topic and its great popularity. After consulting with a great number of sources, in 1584 Irenaeus published a seven-hundred-page volume, *De Monstris*, from the printing press of Nicholas Henricus of Ursel, the publisher of *Image and Mirror*.

The title page of the book describes the seven main subjects of the book: 'Monsters. Concerning Special Wonder-birth ['Wundergeburt']. I. What a monster or wonder-birth is. II. Hundreds of stories of wonder-birth, in chronological order. III. Which one is the most horrible monster. IIII. Who the creator of wonder-birth is. V. What caused them and is their origin. VI. They are punishments for sins. VII. What they mean and what follows later'.[49] At first glance, one receives the impression that the volume is a secular wonder book on monstrous births. The phrase 'Hundreds of stories of wonder-birth' had to be especially appealing to curious readers, and unlike the title page of the *Prognosticon*, there is no clue to indicate the religious nature of this book. However, in fact the latter part of this volume is fully devoted to an attack on the author's theological opponents; yet the polemical nature of this work is not indicated in the title page. Most likely Irenaeus refrained from doing so to promote the book's circulation while avoiding censorship. By the time of its publication, Irenaeus had already become known as a notorious, relentless enemy of the Concord.

But how could Irenaeus connect stories of monsters with his theology? Here the analogy between fallen humankind and monster, first presented in the *Image and Mirror*, shaped the rhetorical foundation of Irenaeus's argument. In the preface of *De Monstris*, he argues that God created human beings in his image, and this means that they were originally good, wise, holy, righteous and obedient to God's will and the Law. However, 'by the abuse of free will' our first parents were deceived by the devil and the Fall took place, which brought 'a gruesome and dreadful spiritual change to humankind'. Now humankind has turned into the disgusting image of the devil or 'the mask of the devil' ('Larue des Teuffels'). Man's whole nature has become evil, spiritually blind, unrighteous and powerless to do good. The soul who turned against God and the evil heart are 'the most terrible monsters and beasts among all creatures, next only to the devil, in God's eyes'. People read in books about many wondrous races with unusual forms and shapes, such as those with dog's heads, with horse or goat feet, without heads, with eyes in the chest area, or with only one eye. 'In our eyes, such people are strange monsters'; however, 'in God's eyes they are not so disgusting and fearful as a godless man who is in sin and blasphemy, alive and living without fear of God'.[50] Thus Irenaeus masterfully links the discourses on monsters and his theology on original sin in a surprisingly persuasive manner.

The gruesome acts of sin and blasphemy committed by human beings from the beginning of history are signs of the monstrosity of godless human beings. In the following pages, various stories of idolatry and superstition are recounted to buttress this claim. Irenaeus displays his new knowledge of human sinfulness by describing the religion of the natives in the New World taken from the accounts of explorations. After narrating how priests in India offer sacrifice to the images of a tiger and peacocks, he recounts that in Peru sometimes God is an animal, such as a bird, tiger or deer. For Irenaeus, these examples from the New World provide convincing material to show what happens if people do not know the true God and remain in original sin.[51]

Many Christians also fell into false worship. Papists brought many errors into the Church, and even among Protestants a number of false prophets appeared and a 'dreadful fall and apostasy from Truth' occurred in his age, due to the new *Jacobandrische[n] Discordien Buch*. After repeating his long-cherished arguments that true prophets are persecuted by the world, he returns to the issues of human vices and narrates various examples of treason, family feud, evil rulers and persecution in the Scripture and history. Here he adds more stories from the New World and recounts the cannibalism of island dwellers named *Carabibern* (Caribbean?). All of these things, 'which man read in the Bible, Church history, and other history, and [we] still daily experience', are 'the fruit and signs of original sin', 'our corrupted nature and being' and 'a bad spring'. Unless we have a new birth, our flesh remains in original sin and will continuously be a source of sins and blasphemy.[52]

At the end of this long preface, the main purpose of this volume is explained. Irenaeus presents this report of misbirths to instruct pious Christians and to warn all people in sin. People should not to view monstrous births with 'epicurean and pig-like eyes' but consider divine punishment in future. God is threatening especially Germany with all kinds of signs of his wrath: 'It is feared that soon it [God's wrath] will come through the Turks or another foreign army, as Luther prophesied'.[53]

This preface is a remarkable piece. Irenaeus's connection of people's interest in monsters with his theological stance, based on the analogy between the monstrosity of corrupt human nature and monstrous races, shows the outstanding creativity of his writing. Though the works he consulted have not been identified, he even exploits the latest accounts of New World explorations to support his understanding of corrupt humankind. However, the monsters he mentions in the preface are mainly the monstrous human races in foreign lands that often appeared in medieval literature. How can he connect his theology with contemporary news of monstrous births as described on the title page? In the main text, he transforms the stories of deformed children into religious propaganda.

De Monstris: The Main Text

In the first chapter of the main text, Irenaeus discusses the definition of monsters based on ancient authorities. Chapters 2 and 3 are catalogues of *Wundergeburt* stories, and they had to be most appealing parts for common readers. For 140 pages, numerous cases of deformed births that occurred between the time of Herodotus and the year 1581 are narrated. Irenaeus indicates the sources of his stories, which include books by Conrad Lyscothenes (the major source), Fincel, Goltwurm, Kaspar Peucer, Sebastian Frank, Luther, Melanchthon, Erasmus, C. Spangenberg, Simon Pauli and Achilles Gasparus, as well as the chronicles of various places. The sources for cases between 1570 and 1581 are not given, probably because they were directly taken from contemporary broadsheets and tracts. His list of sources indicates that even in exile Irenaeus did not lose his intellectual curiosity and continued his collection of stories of wonder.

The cases in these chapters include the news of the androgynous monster of Ravenna, a *cause célèbre* of 1517,[54] and the story of the Forest Demon ('Forestteuffel'), a monstrous creature captured near Salzburg in 1531, whose mysterious statue we can still see at the Hellbrunn Castle.[55] One of the most shocking examples of deformed birth in this catalogue is that of a baby born in a place named Chiory in January 1578. The child, who was 'rather a devil or evil spirit', had a huge head with large ears and three horns, together with a long tail that hung from the head and consisted of red flesh. The colour of the baby's body was red and black, like the colour of coagulated blood. The hands were 'like a devil's claws'. Since Irenaeus states that an image of the baby was printed in Prague, most likely he obtained this information from a broadsheet.[56]

In the following chapters, Irenaeus argues that these poor deformed children were created not by the devil but only by God. But why does God create such children? According to Irenaeus, the function of a *Wundergeburt* is God's punishment and sign, a reminder of the Fall and our corrupted nature. Monstrous birth is also a special *Merckzeichen* (mark) that God shows to the world to press people to repent. An interesting moral tale from Spangenberg's catechism is quoted here: there was a noble who went hunting every Sunday, and his wife delivered a baby with the head of a dog. Another example is of a *Junker* in Thüringen who beat a pastor severely. The *Junker*'s wife gave birth to two children, but neither of them had any bones in their bodies. He regretted what he did to the pastor, and once the pastor forgave him, his wife no longer delivered deformed babies. Irenaeus also warns that God often punishes people's arrogance and luxurious lifestyle through the birth of babies with slits in their bodies, like the slits in coats and trousers; a strange baby born in the Marck in 1550 had bit slits on its legs.[57]

These stories of divine punishment may have been morally interesting, but Irenaeus develops his arguments further because his aim is to connect *Wundergeburt* with his polemical agenda. His next step, in Chapter 7, is to connect the deformities of babies with the condition of this world. For example, in 1525 a headless child was born in Wittenberg; that year, with the outbreak of the Peasants' War and the death of Elector Friedrich, Saxony fell into a condition that 'lacked a head'. There were monstrous births in 1531, and Irenaeus claims that these babies' deformities signified the true nature of papist theologians who attack the Augsburg Confession. In several cases monstrous births can also be connected to wars: an example is Melanchthon's story about the Franconian baby with 'a knife on the stomach', which was regarded as a portent of the Schmalkaldic War.[58]

All of these tales signify the connection between monstrous births and the secular order; but as we know, Irenaeus's primary concern is their relation to the condition of the Lutheran Church. In 1564 a baby with a sharpened head and four bumps on the body was born, and this news was recorded by Fincel. Irenaeus quotes Fincel's Latin interpretation and adds his own German translation. The 'sharpened head' signifies the sinfulness of human reason: although reason is clever, it is blind to spiritual matters. The four bumps mean four prominent errors, such as the *Schwarm der freywilligen Herrn* and *Zwinglianer oder Calunister Schwarm*. Understandably, to this line Irenaeus adds that 'now one may say *Accidentzer Schwarm* too'.[59]

After some discussion of half-animal babies (which do not relate to religious matters), Irenaeus provides two sections that directly connect misbirths and his attacks on his theological opponents. The title of the first section is 'What the double-tongued monster ['Zweyzüngige Monstra'] means'. Irenaeus open this section with the statement that the double-tongued monster generally means 'double-tongued, changeable, and slippery teachers or other people whose mouths do not have conscience and consistency'. During the time of Constantine, there was a child born with two mouths, two tongues, four eyes and short ears. Such a *Wundergeburt* could foretell many things, but in this case the child signified the double-tongued followers of Arius because while they confess that Christ is the Son of God, in their hearts they deny it. Irenaeus points out that births of similar children took place several times during recent years. In 1551 a child with a black body was born in Tannewalde. The baby had two horns on the head, and its large mouth had an unusual tongue – it was square, and its colour was white. A baby born in Lüneburg in 1553 had a strange mouth with a tongue the length of a finger, divided in two. A similar birth also took place in the same city in 1550. In 1559 a child with a large mouth and a divided tongue was born in Aschersleben (Irenaeus had served as a deacon in Aschersleben until 1559 – did young Irenaeus see the baby with his own eyes?).[60] Why do the births of babies with strange tongues occur so frequently? Irenaeus offers readers his answer in the next section.

The title of the next section is 'Philip Melanchthon's double-tongued teaching on Holy Supper'. Why is Melanchthon's teaching concerning the Holy Supper related to the deformity of babies? For Irenaeus, there was an important connection. He claims that while Luther was alive, Melanchthon understood the nature of the Holy Supper correctly and confessed it. However, after Luther's death 'one different, unpleasant tongue was found' in his mouth and he left his past position for a Sacramentarian understanding of the Holy Supper. This double-tongued mouth of the German Preceptor is what double-tongued monsters mean, Irenaeus declares.[61] But why can we connect Melanchthon with the *Wundergeburt* cases of Lüneburg? Here Irenaeus reveals the source of his information. The two cases (1550 and 1553) were recorded by a Lüneburg preacher, Antonius Wipperman, who reported these cases at the end of his booklet against Georg Major's controversial teaching on good works. One day a child was taken to Wipperman to be baptized, but when he saw the face of the child, he was shocked because from the baby's large mouth a finger-long tongue stuck out and was divided in two. A similar incident occurred during the time when the Church was in conflict over adiaphora. Wipperman states that 'without doubt' these cases meant the danger of being double-tongued with neutrality over religious matters in the Church.[62] Melanchthon was the one who employed the idea of adiaphora; Georg Major was Melanchthon's close friend, and together they formulated the Leipzig Interim. It is no wonder that Irenaeus saw a connection between Melanchthon and the double-tongued deformities in Lüneburg in the 1550s. The accusation of crypto-Calvinism against Melanchthon is a new element in Irenaeus's wonder books, but considering the expanding influence of Calvinism in Germany and the fact that many crypto-Calvinist ministers left the Lutheran Church, it is not difficult to imagine why Irenaeus chose Melanchthon as his first target.

At this point the direction of Irenaeus's focus suddenly changes. After his invocation of God's help, the *Zweyzüngigkeit* (double-tongueness) of false teachers in the Lutheran Church becomes the main topic of Irenaeus's arguments, and he discusses this matter throughout the rest of the book. Discussions about *Wundergeburt* have been totally forgotten until almost the end. The theologians who became targets of Irenaeus's attack include Major, Strigel, 'young theologians of Wittenberg', Tileman Hesshus, Wigand, Mencel, Georg Autumnus, Hieronymus Peristeri and of course his arch-enemy, Andreae. Irenaeus quotes his opponents' writings concerning the doctrine of original sin or the Concord and accuses them of having changed their convictions. For example, against Mencel, Irenaeus points out that when a colloquium was held at Weimar in 1571 to solve the controversy, Mencel declared that he and the preachers of Mansfeld confessed human nature was sin itself. In Irenaeus's eyes, Mencel had clearly once agreed to the Flacian understanding of original sin.[63] However, not long afterwards Mencel got a 'gruesome, disgusting *Mameluckische* tongue' besides the 'Lutheran tongue', and soon he abandoned his original position.

The greatest target of his attack is 'Doktor Jacob', for whom Irenaeus devotes forty-four pages to contradict his theological positions. It was not a difficult task to find inconsistencies in Andreae's writings, as he often changed approaches and tactics to achieve the Concord. Andreae's success was largely owed to this pragmatic attitude, but for Irenaeus such an attitude was nothing but proof of Andreae's *Zweyzüngige* character. For example, Irenaeus points out that Andreae had once said that the teachings of the Lutheran Church for these thirty years had been fundamentally in unity, and all theological controversies in the Lutheran Church were only over questions of wording. Irenaeus ridicules this claim, observing that Andreae's opinions themselves were fundamentally not in unity![64] When Andreae came to Tübingen, he even signed Strigel's confession with his own hand.[65] One by one, Irenaeus points out how Andreae had been lukewarm towards the various theological errors of Osiander, crypto-Calvinists and others. In this way, Irenaeus accused prominent theologians of their 'double-tongued' inconsistency. To some degree Irenaeus's claim was valid: many of these theologians had at first been favourable towards Flacius's view of original sin, and the theologians of Jena University had shown a strongly negative attitude towards Andreae's project of the Concord. It is not strange that Irenaeus perceived the changes in their convictions as a sign of their inconsistency and shallow faith. Thus Irenaeus struck these theologians at their weakest point.

Close to the end of the book, Irenaeus returns to his discourses about monsters as if there was no diversion, but his polemical tone remains. He discusses 'what the children with two, three or more heads mean' and interprets that such babies signify disunity in religious and secular *Regiment*. The two-headed baby of Weitersweiler in 1511 was a sign that foretold the great split within the Church.[66] There were a number of births of two-headed babies in the 1540s in Germany, and these were signs foretelling the outbreak of the Schmalkaldic War.[67] Then Irenaeus changes the direction of his argument. The '*biceps Monstrum* or two-headed child' born in Worms in 1578 (the case mentioned in the *Erklerung*) signified that a 'great split and division in religious and secular Regiment are at hand'. This is related to Andreae's Formula. Here Irenaeus repeats the familiar arguments that God's Word shows that within the Church, disunity, splits and conflict will continue until the Day of Judgement 'like similar *bicipipitia monstra* or two-headed wonder-births'.[68] The rest of his argument, which continues for another ten pages, does not differ much from his criticism of the Concord in the *Prognosticon*: no righteous, eternal unity can be expected between Christ and Belial.[69]

The structure of *De Monstris* and that of the *Prognosticon* are similar. In both cases, the first half looks like a regular wonder book: a catalogue of stories of wonder is presented, and a moralistic Christian exposition explains what readers should learn from these strange phenomena. After these stories combining *Wunderzeichen* with Irenaeus's polemical agenda, the arguments abruptly turn to theological discussions over false teachings. In both books, the connection between the first half of the book and the latter half is awkward. Indeed, in *De Monstris* Irenaeus is obviously making an effort to reconnect the two topics at the end of the volume. However, he probably did not feel the weakness of the connection between the two parts of his books because for him, there was a tight relationship between *Wunderzeichen*, theological errors and the coming of the Last Judgement. Thus in the *Prognosticon*, the *Wasser-Spiegel* and the *Abdruck*, Christ's prophecy of the End Times serves to unite Irenaeus's complicated argument at the conclusion.

The stories of wonder catalogued in these books had to draw the attention of many, including those who did not have much interest in theological controversies. The phrase 'hundreds of stories of *Wundergeburt*' that Irenaeus used on the title page suggests his intention to attract as many people as possible. Irenaeus could expect that whether they shared his theological understanding or not, those who purchased his books would also be exposed to his religious polemics. Irenaeus's connection of two-headed babies with disunity in the Church and the secular *Regiment* may have been accepted by many readers because it was common to interpret the deformity of babies as a reflection of the condition of the world. However, was Irenaeus's connection of double-tongued babies with the inconsistency of Lutheran theologians really persuasive? This is questionable, because he quotes only four or five cases of babies with this type of deformity, while he names many theologians as *Zweyzüngige* teachers. Actually, in his argument Irenaeus connected only Melanchthon's inconsistency directly with the Lüneburg cases. However, Irenaeus regards *Zweyzüngige* babies as emblems of the double-tongued nature of all his opponents and the divine judgement on them, and he skilfully used the stories of the babies to lead readers to his theological view. In this sense, the rhetoric of double-tongueness worked.

Conclusion: Did Irenaeus's Strategy Succeed?

Irenaeus's dramatic experiences between 1572 and 1574 as well as the success of Andreae's Concord project had to strengthen his sense of being a 'remnant' of the true believers, and this forged his identity as an *Exul Christi*. Probably trained by his underground writing and intense publication activity, in his next wonder book, *Prognosticon*, he established his strategy to draw readers' attention by the eye-catching title page and stories of wonder, while hiding his polemical agendas. Despite his difficult life in exile, he kept collecting stories of wonder from various sources, especially those related to monstrous births, including an account of the exploration of the New World. This effort was crystallized in *De Monstris*. In this book, he skilfully connects his theological conviction with the monster discourses of his day by pointing out the parallel between the shocking outlooks of monsters and the depravity of human nature brought by the Fall. Then in the following chapters he attracts readers with hundreds of stories of misbirths and stresses that babies' deformities reflect the confused condition of the world, a view widely shared in early modern Europe. Then he suddenly connects *Zweyzüngige* babies with the inconsistency of his opponents and leads readers to his thundering polemics.

It may not be fair to argue that for Irenaeus, *Wunderzeichen* served only as bait to draw people to his theological agendas. It is obvious that he had a genuine interest in stories of wonder, and there is no reason to doubt his passionate conviction that wondrous phenomena reflected God's wrath on human sins, especially theological errors. Indeed, as Wipperman's account of the Lüneburg baby shows, Irenaeus was not the only figure who connected double-tongued deformity with unfaithful inconsistency among contemporary theologians. Here we must consider the four leitmotifs that repeatedly appear in his *Wunderzeichen* writings: God's signs of wrath, the prevalence of false teachers, the persecution of believers and the coming of the Last Judgement. In Irenaeus's world view, these topics are tightly intertwined and inseparable because Christ himself states that these things will appear prior to the End Times. Irenaeus himself experienced Christ's prophecy being fulfilled one by one – through his witnessing of a dramatic *Wunderzeichen*, his struggles against false teachers and his suffering as an *Exul Christi*. For him, contemporary controversies over true doctrine had cosmological significance, and there was nothing strange in a God who sent signs to warn people not to be deceived by false prophets. Though in modern eyes his combination of wonder book and religious polemics seems awkward, Irenaeus and probably many of his contemporary readers had no reason to feel so.

One question remains: how did contemporary people react to Irenaeus's wonder books? This question is not easy to answer. Though several theologians published refutations of Irenaeus's works, none of them discusses his *Wunderzeichen* literature. Thus it is difficult to measure whether Irenaeus's strategy to influence readers' theological understanding by *Wunderzeichen* worked as he had hoped. However, we do know that his works on wonders acquired a number of readers despite his unorthodox theological positions. In 1595 his *Wasser-Spiegel* was reprinted by the Eisleben printer Bartholomaeus Hörnig, commissioned by the prominent Leipzig bookseller Henning Gross. It is not clear what motivated Gross to reprint this thirty-year-old text, but there were several possible factors that made this work alluring at that timing. The severe winter of 1594 was followed by a sudden improvement of temperature in late January 1595, and the melting snow caused floods in the Saale river. Further melting snow, ice drift and rain brought a series of serious inundations throughout early March.[70] Moreover, Leipzig was still recovering from the trauma brought about by the failed attempt of the 'Second Reformation' by Elector Christian I and his councillors, and hostile sentiments against the Calvinists were still running high. The anti-Calvinist riot in the spring of 1593 ended with the bloody beheading of four men, and the citizens of Leipzig showed their sympathy with those executed by burying them honourably – an unusual treatment for executed criminals.[71] It is no wonder that a shrewd businessman like Gross regarded the *Wasser-Spiegel*, which discusses both floods and Calvinist evils, as timely material to publish. In addition, this book was a 'safe' work among Irenaeus's writings, which attacked mainly the Calvinists and the Synergists, both condemned by the Formula of Concord.

Indeed, evidence suggests that the *Wasser-Spiegel* was cherished as a rare book focused on watery disasters. In 1615 the Kitzingen pastor Johann Aldenberger published a lengthy book titled *Mirror of Fire, Water and Wine* (*Fewer Wasser vnd Wein Spiegel*), which catalogues fires, floods and the abuse of wine in history up to the year 1614. In the preface, the fifty-three-year-old pastor points out the necessity to update the history of floods. He states that since Irenaeus published his book in Eisleben forty-eight years ago few *Historien* can be found on the topic, and 'according to my knowledge, it [cataloguing of floods] has not been continued for more than 50 years'. Encouraged by the 'request and wish of some of my good friends and local people', he compiled this catalogue as a mirror and a Christian reminder that sins and blasphemy would bring divine punishment, as happened to the First World.[72] In the beginning of the *Wasser Spiegel* section, Aldenberger quotes several passages from Irenaeus's book regarding God's watery punishments on epicurean sinners.[73] It is not clear whether Aldenberger consulted with the original 1566 edition or Gross's 1595 edition, but his comment indicates that there was still demand of not only for Irenaeus's book but also for a subsequent volume. Actually, the bookseller Gross did not miss this chance again – Aldenberger's book was published in Leipzig by Gross.

It seems that Georg Rost, court preacher of Lübz, was just such a reader looking for a book on floods. Inspired by the dreadful storm and inundations in February 1625, he published *Double Theological Mirror* (*Zweyfacher Theologischer Spiegel*), which consisted of *Thewrungs Spiegel* and *Wasser Spiegel*, through a Rostock publisher. In the latter part, after discussing the fact that storms and floods are God's punishment on sins and a harbinger of Judgement Day, he recounts that history shows how various disasters follow such inundations. In the section on 'how flood of water is followed by death and pestilence', Rost quotes Irenaeus's report of the flood in 1345 and the subsequent plague in Europe.[74] Almost at the end of the book, he provides a section on 'how flood of water is also followed by heresy' and presents an Irenaesque argument on the parallel between floods and theological errors: 'the sea often spreads and floods many cities and lands. Heresy and false teaching also flood multiple land[s] and multiple cit[ies] with God's approval, because human beings enjoy lies more than truth, as Paul says'. His subsequent reference to Revelation 8, in which the Beast comes out of water, heightens the apocalyptic atmosphere. Though the Synergists do not appear, Rost's list of false teachers includes not only the Calvinists but also the Jesuits and the followers of Valentin Weigel and Johann Arndt.[75] Did he get inspirations of these passages from Irenaeus's discussions? Though his writing style is less creative than Irenaeus's, his arguments suggest that Irenaeus's rhetoric, which combines natural phenomena, apocalypticism and false teachers, was still usable in the theological controversies of the seventeenth century.

These references to Irenaeus's *Wasser-Spiegel* do not mean that his other wonder books were not widely read. In his comet chronicle (1579), Georg Caesius briefly mentions the *Prognosticon* as one of the sources of his book.[76] This shows that Irenaeus's comet book reached the hands of the pastor of Leutershausen within one year of its publication. The next reference to the *Prognosticon* appeared almost forty years later. Shortly after the great comet appeared in the sky in November 1618, Aaron Bruckhart, preacher of St Ulrich Church in Magdeburg, published a pedantic sermon concerning the ominous comet. At the end of this work, Bruckhart attaches a list of books that can be used for further references. Together with works by Andreas Celichius, Caspar Peucer and Andreas Northius, 'Prognosticon de Comet. Christophori Irenei' is also mentioned.[77] It may be surprising that the pastor of a prestigious church in Magdeburg recommended such a controversial work to the public, but Bruckhart probably believed that the meanings of Irenaeus's polemics in the *Prognosticon* had been lost and it would no longer pose a serious danger to Lutheran orthodoxy.

Though fragmentary, these sources suggest that Irenaeus's wonder books were circulated widely. Aldenberger, Caesius and Bruckhart regarded Irenaeus's works as books on wonders and disasters, not polemical pieces. This might not have been Irenaeus's true intention, but it tells us that at least his strategy of catching readers' eyes through stories of wonder succeeded. Though Rost mentioned Irenaeus's name only as a source of information, his reading of the storms as signs of the emergence of heretical teachers may reflect Irenaeus's attack on the Synergists and the Calvinists. If so, Irenaeus's efforts were not in vain: he at least inspired a minister to defend people from theological errors and to warn against the danger of false teachers. As Irenaeus's works show, celestial wonders, confessional conflicts and apocalypticism were closely interconnected in the cosmology of late sixteenth-century Lutheranism, and it was still relevant to the people living during the dark days of the Thirty Years' War.

6 ANDREAE'S PASTORAL USE OF *WUNDERZEICHEN*

In the previous three chapters, we have seen examples of how *Wunderzeichen* were used as polemical weapons. In this chapter, we will see that these stories of wonder could also be used for pastoral purposes. The protagonist of this chapter is Jacob Andreae, Irenaeus's deadly foe. Andreae did discuss *Wunderzeichen* in some of his works, but when we read these discourses, we receive the impression that he was hardly enthusiastic about celestial phenomena. In this aspect, he showed a marked difference from his co-formulators of the Concord. We have seen that Selnecker published a prayer booklet on the 1577 comet, while Chytraeus compiled a tract about the 1572 nova and the 1577 comet. Musculus's interest in celestial wonders is obvious from the repeated references to *Wunderzeichen* in his writings. As for Chemnitz, his passion in astrology was not mere youthful dreaming. In 1576, when the University of Helmstedt was founded, Chemnitz advised his lord, Duke Julius of Braunschweig-Wolfenbüttel, the founder of the university, to 'give careful attention to the arrangement of the sky, to ascertain when there would be a favourable constellation'. The realistic duke had to remind him that for the future of the university, it was more important to give careful attention to finance rather than to look at horoscopes.[1] Such an episode could hardly occur in the life of Andreae, who did not hide his suspicion of astrology.

Andreae mentioned celestial phenomena in his works only for practical reasons – to make his messages more attractive and comprehensible for the laity. In other words, he quoted stories of wonder in aid of his pastoral concerns, while celestial signs themselves were not at the centre of his interest. Indeed, pastoral concern was the driving force in Andreae's active career. According to Robert Kolb, although Andreae served earnestly as the provost of Tübingen University, a ducal counsellor of the court of Wurttemberg and a formulator and promoter of the Concord, he 'was not so much a theologian as a parish pastor and a popularizer of the thought of others'. Even after he left his pastoral position at Göppingen to become a university professor, 'concern for the pastoral implications of Evangelical doctrine dominated much of his thinking and writing'. Andreae's theological training at Tübingen University was rather shallow, and his theology was 'neither deep nor complicated'. Nonetheless, as Kolb stresses, Andreae poured his energy into teaching the principles of Lutheran faith to laymen so that they could explain their faith and defend it from the teachings of the papists and the Calvinists. Andreae employed four methods in order to achieve these goals: giving answers to questions or feelings raised by the laity; appealing to people's common sense; constructing arguments based on the catechism 'the fountain that cannot be drunk dry'; and supporting arguments with Scripture.[2]

In this chapter, I examine three works by Andreae that mention *Wunderzeichen*, and I analyse how he used popular interest in wonders for pastoral guidance. My analysis supports Kolb's observations, as these texts reveal Andreae's two primary concerns: how to make his message more comprehensible to the laity, and how to rebuke the errors of the epicureans so that they do not become *Ergernussen* (stumbling blocks) for other members of the Church. Here the term 'epicureans' is used to mean those who misunderstand the Gospel to claim that since Christ died for humankind's redemption, they are free from all Christian discipline. Andreae worried that such an immoral attitude gave the Anabaptists and the spiritualists an opportunity to attack the Lutheran Church as a false church which did not bring moral regeneration in believers' hearts.

The first text to be examined is Andreae's exposition of Christ's prophecy of the end of the world in Luke 21, published in the last years of Andreae's life. Here he mentions the strange celestial phenomena of his days, but he shows a subtle difference in tone from the other theologians we met in Chapter 2. The second text was written in response to the great eclipse of April 1567, and aimed to turn people's attention to God's Word rather than vain astrological predictions. However, interestingly, in order to draw people's interest to his messages, he himself used the popular astrological imagery of the planets. The last text is the lengthy sermon Andreae delivered in Weimar shortly after the appearance of the great comet of 1577. The main goal of this sermon is to teach the laity about orthodox Lutheran doctrines (according to the Concord), and how they differ from erroneous teachings such as the Flacian understanding of original sin. In addition, he places the rise of the 'epicureans' of his day within the history of the Christian Church, interpreting this as a sign of the End Times. The comet is mentioned only briefly as the sign of the imminent coming of the Last Judgement, showing God's grace for true believers while threatening the godless with his wrath. The text amply shows how Andreae's four methods of approach, explained in Kolb's article, worked to indoctrinate the audience.

The examination of these texts reveals a strong contrast between the pastor Andreae and the polemicist Irenaeus. While both detested astrology, they also used astrological imagery to draw people's attention (in Irenaeus's case, by giving the title *Prognosticon* to his comet book). It seems that Andreae was reluctant to quote *Wunderzeichen* stories in his messages, possibly because he feared that references to celestial phenomena would encourage people's interest in astrology. Irenaeus never considered such a possibility and did his best to turn people's attention to celestial phenomena. This illuminates Andreae's thoughtful pastoral consideration for his people.

There is a growing trend among recent historians to see Lutheran preaching during the late Reformation era in a more favourable light than traditional views. For example, Gerald Strauss's view that the main goal of pastors' catechism was not to teach the theological doctrines themselves but to create obedient subjects for princes is criticized by Mary Jane Haemig, who insists that these pastors made serious efforts to educate the people through catechism so that they could defend their theological understanding from dissenters.[3] Practical pastoral care was never forgotten, as the great number of sermons and devotional literature published for consolation of those in distress indicate.[4] Preachers also tried to incorporate the laity's languages and daily concerns into their messages to make them more acceptable.[5] The contents of Andreae's messages support these views, and his use of celestial phenomena shows his conscious effort to make his messages more comprehensible to common hearers.

Andreae's Exposition of Christ's Apocalyptic Prophecy

Unlike other prominent theologians we saw in Chapter 2, Andreae never collected his sermons into one single postil. However, he occasionally published his pericopic sermons, and we know about his view on Christ's prophecy about the End Times through his *Four Christian Sermons on the Gospel of Advent* (*Vier Christliche Predigten Vber etliche Euangelia im Aduent*; 1588). In the section of the second Sunday of Advent, he focuses on Luke 21 and reminds readers that 'great, dreadful signs were seen in the sky in the past years'. Many comets appeared one after the other, and especially noteworthy was the 'gruesome dreadful comet' of 1577, which stood in the sky for a long time 'like a fiery rod'. The comet preached to the godless about the dreadful justice and punishment of God that the world would undergo. 'In addition', Andreae states, 'how many times ... in all places in Germany ... the whole sky looked burning?' Many people could only think that the Day of the Lord was at hand. Yet these signs were soon forgotten 'like a drop in the water' as if no one saw such a sight, although God is really warning of the danger of sudden punishment and eternal damnation through these signs.[6]

At a glance, there is not much difference between Andreae's exposition of apocalyptic signs and that of the other theologians we saw in Chapter 2. However, Andreae's listing of *Wunderzeichen* is much simpler than that of other theologians. He does not catalogue specific examples of strange phenomena, and the only example he mentions is the great comet of 1577, witnessed by everyone.[7] He devotes far more energy to explaining the next apocalyptic danger:

the spread of vice among people (Luke 21:34). He spends one and a half pages on celestial signs, but he uses nine pages to discuss the vices of gluttony, heavy drinking, *Sicherheit* and 'worries of life', while also attacking people's extravagant lifestyles that waste so much money. Who can doubt that the Last Judgement is approaching if such living is the general course of the whole world? Andreae concludes that all the prophecies of Christ have been fulfilled and God is now calling us to repentance and 'a sober, modest Christian life and way' so that we will not suffer eternal damnation on the last day.[8] After explaining why the Last Judgement is a joyful occasion for true believers and how wonderful the time of salvation is, Andreae returns to the issue of the 'worries of life' and warns of the dangerous trap of human greed. This topic, which continues for ten pages, offers him an opportunity to attack the practice of usury. He devotes the next seventy-two pages to this discussion.

In this exposition, it is clear that Andreae shows much more interest in the necessity of people's moral improvement than in celestial phenomena. Of course, we cannot measure Andreae's interest in *Wunderzeichen* solely based on this single source. As we saw in Chapter 5, Irenaeus discusses celestial signs only briefly in his exposition on Luke 21, while he had lifelong interest in stories of wonder. The spread of vice and *Sicherheit* was part of Christ's prophecy of apocalyptic signs, and many theologians discussed the topic in their postils along with their expositions of celestial wonders. However, Andreae's message suggests that his focus was more on practical issues in the real lives of people rather than on signs in the sky. The next source, written almost two decades before this sermon, shows why Andreae was more interested in the moral problems of the age than celestial phenomena.

Christian, Necessary and Faithful Reminder (1568)

On 19 April 1567 a great eclipse was observed in the sun and surprised many people because 'for many years such a thing had not occurred'.[9] Andreae knew what would come next: a flood of astrological tracts discussing the meaning of the eclipse, filled with irresponsible predictions embellished with sensational words. By the end of the year, he had completed his unique book, *Christian, Necessary and Faithful Reminder, Presented According to the Course of the Terrestrial Planets* (*Christliche notwendige vnd ernstliche Erinnerung Nach dem Lauff der jrrdischen Planeten gestalt*; 1568), to admonish people not to put much faith in astrology but to focus on God's commandment.[10] Consisting of one long introduction and five sermons, this text insists that if people want to escape from misfortunes, they need moral regeneration and Christian discipline rather than unreliable astrological predictions.

What kind of astrological prediction was on Andreae's mind? He expected that many natural philosophers would 'especially write and predict about many misfortunes'.[11] We have several examples of *prognostica* on the 1567 eclipse that predicted numerous imminent calamities, as Andreae expected. Paul Severus believed that inflation and wars would strike Alsace, Lombardy, Spain, Italy and Bavaria. In Germany, many churches and cloisters would be devastated and many clerics would suffer persecution, while 'a great, mighty folk' (obviously the Turks) would invade Hungary, Austria, Bohemia, Meissen and Hessen. A new persecutor of Christianity would appear in the Netherlands and torment believers with sword and fire, but by God's revenge he will have a dreadful end.[12] Such stormy conditions would continue until 1570, but after the year 'a new Reformation both in the spiritual and secular [realms] will take place' and good order and law would be established. A pious emperor would rule the empire and restore Christianity in Hungary and Constantinople.[13] This type of optimistic prophecy of the golden age under a pious emperor was the familiar stock of medieval prophecy, but it was certainly not appealing to a pragmatic diplomat like Andreae, who knew the world of princely politics well. The predictions offered by Bartholemeus Scultetus, a mathematician of Görlitz, were more sophisticated and realistic: in his astrological tract, Scultetus argues that the 1567 eclipse foretold the spread of errors and the split in faith because it signified that the eternal sun, the light of God's truth, will be darkened.[14] Andreae had to share Scultetus's concern about the split of the Church, but he was much more interested in how to heal the split than in fanning the anxiety of the people.

In the introduction/dedication of the *Reminder*, Andreae explains why he wrote the book and what he wants to achieve. First, he opens with a brief report about the great eclipse in April. Andreae reminds readers that astrological predictions are not always reliable because even the difference of one minute can change the result. He had once heard from a prominent mathematician that providing an astrological prediction about a specific person and land can be comparable to throwing dice! Since human nature is filled with curiosity, we always want to know ahead of time how the climate of the future is, whether there will be an epidemic, and whether peace will be preserved. However, God reprimands such curiosity and trust in astrological art. Actually, writes Andreae, people already have 'an unchangeable, clear, and infallible *Practick*' which is so simple that it can be called 'Farmer's *Practick*' or 'Farmer's Calendar'. This tells people what they should do to avoid misfortunes, without many complicated astrological calculations. It is the twenty-eighth chapter of Deuteronomy, which describes God's promises that the people of Israel can enjoy peace, good climate, good health and abundant harvest as long as they obey his commandments. However, if they do not obey God's law, people will be cursed and tormented by pestilence, drought and war. 'This is a fine, simple *Practick* and general Farmer's Calendar', Andreae states, 'which everyone knows ... and can easily judge by' whether we can expect 'good or bad weather, peace or war, health or sickness, inflation or fair price, freedom or bondage'.[15] Unlike ordinary human *Practicken*, which may contain many wrong speculations, this *Practick* is perfect and can be applied to any year.

At this point Andreae turns to Christ's prophecy of the End Times and the moral failures of the people. Christ foretold that signs would appear not only in the sun and the moon but also on earth, namely the spread of vice among the people. Seeing that vices are prevalent among all classes of society, we should 'calculate' that not just a temporary downfall of our fatherland but the Last Judgement itself is approaching. This topic leads Andreae to consider the problem of Christian discipline. Though people were now liberated from the abuse of the papists, this caused the collapse of Christian discipline, and many people were now living an 'epicurean, wild, beastly life'. Andreae insists that we should not discard the discipline of Christian living along with the idolatry of the papists. These epicureans withdraw not only from papist practices but also from the true teaching of the Gospel and thus become enemies of 'true fear of God, Christian discipline, and respectability'. This moral failure of the people raises another concern for Andreae. The Anabaptists and the Schwenckfeldians attack the lack of moral regeneration among the Lutherans and claim that this failure is a sign that the Lutheran Church is not the true church. Andreae states that these *Rottengeister* (corrupt spirits) are confusing people by suggesting that since there are so many *Ergernussen* (stumbling blocks, in this case moral failures) in the Lutheran Church that Lutheran teachings cannot be true. Though he immediately rejects such claims with several questions (Is Christ's teaching false because there was Judas among the disciples? Were the messages of Isaiah, Jeremiah, Hosea, Amos and other prophets false because people did not obey them?), he admits that the 'sinful, stumbling, disorderly and blasphemous lives' of people are contributing to these attacks. Thus calling the German people – who are indulging in all kinds of vices in the name of the Gospel – to sincere repentance and conversion is the aim of Andreae's book. He insists that in order to achieve this goal and restore the discipline of Christian living, it is important that ecclesiastical and secular authorities collaborate with and support each other. Andreae reminds readers that they are threatened by the dangers of inflation, plague and the Turks, and he states that it is time for all classes of people to repent and seek God's grace, like the people of Nineveh.[16]

This introductory chapter, written as a dedication to Count Ulrich von Helffenstein, clearly shows Andreae's main concerns and his rhetorical skill as a preacher. He opens with a reference to a recent celestial wonder, an eye-catching topic. While he immediately rejects the prevailing astrological predictions as products of an unreliable art, he gives a biblical *Geistliche Practick* which admonishes people to obey 'God the Lord's Church and Police Ordinance' ('Herrn Gottes Kirchen vnd Policeyordnung'). Then he turns to Christ's prophecy of the End Times, a common topic in *Wunderzeichen* discourse, and guides readers to the problems of moral failures and the 'epicurean' antinomian attitude of some Lutherans. The flow of his arguments is natural, and readers are easily guided from a discussion about celestial phenomenon and astrology to one about the moral problems plaguing the Church. His attack on Anabaptists and other groups and his call for collaboration between ecclesiastical authorities and secular authorities also show his practical attitude.

The main section of the *Reminder* again shows Andreae's ability to exploit the popularity of *Wunderzeichen* news and astrology and to use celestial imagery to catch readers' attention. It is divided into five sermons on Christ's prophecy of the end in Luke 21, and he explains the structure of this collection in the first sermon (intended for the second Sunday of Advent, according to the author). Christ foretells that three kinds of signs will appear prior to the Last Judgement: signs in the sun and the moon; signs in the water; and signs on earth and in human beings. 'Since we have often heard about the first two signs', Andreae states, we shall focus on the last category. Christ warns us to be on guard and not weigh our hearts with gluttony, heavy drinking and worries of life (Luke 21:34) so that we will not be surprised on the Day of the Judgement. Be alert at all times and pray, as Christ admonishes. 'With these short words, Christ the Lord reminds us of the five terrestrial planets' that are shining over the earthly kingdom and ruling the world: heavy drinking; gluttony; worries of life (especially greed); false sense of security; and neglect of prayer. Although the Gospel is taught throughout Germany, these vices prevail everywhere, and it is easy to predict that not only will the Last Judgement come but also that a gruesome, dreadful punishment will soon befall ungracious Germany, most likely executed by the Turks.[17] Since the Turkish invasion of Austria in 1566 shocked Germany and renewed people's fear, this was not an empty threat. Yet Andreae urges readers not to despair because God has no pleasure in the death of sinners (Ezekiel 18:23). We must repent and amend our lives.

After this general introduction, Andreae explains the vices symbolized by the five earthly planets and emphasizes how hideous these evil deeds are in God' sight. For example, God hates heavy drinking because, once intoxicated, a person forgets the knowledge of God and becomes a prisoner of the devil. Andreae associates gluttony with people's luxurious clothing and laments the excessive spending of his age; no one eats meals and wears clothes appropriate for his or her class. According to Andreae, such extravagant spending was not an original custom of Germany but recently brought in by Italian merchants. He reminds readers that God's will was manifested to the Germans prior to other peoples, and he denounces their ingratitude towards God's grace. Many people continue with their sinful lives, while believing that 'we have learned that only through faith in Jesus Christ we are saved and by his death our sins are forgiven. We know that we do not have to pay [for salvation] through fasting, masses, prayer and other works. Therefore, let us be free from these works'. Indeed, our sins are forgiven only through the blood of Christ, but it is not true faith to believe that since we have been saved we can therefore indulge in vices.[18]

These calls for repentance are fairly conventional, but Andreae's comparison of the five apocalyptic vices with the five planets shining over the world is unique. While he did not trust astrology, he exploits popular astrological imagery to make his argument more understandable and eye-catching to common people. Andreae identifies the root of epicurean living with the misunderstanding (or deliberate distortion) of Christian freedom, and he patiently admonishes that justification by faith does not free one from Christian discipline. At the same time, he states that many vices did not originate in Germany but were brought in by Italian merchants, possibly trying to appeal to people's patriotic pride by attributing vices to foreign origin.[19] These points indicate that he was a skilled preacher and pastor who made sincere efforts to communicate with the laity. The evil of epicurean living and the necessity of moral regeneration remained chief concerns for Andreae, and he repeats the same topics in a sermon delivered nine years later.

The Weimar Sermon on the Unity of Lutheran Faith (1577)

On 24 November 1577, only two weeks after the great comet appeared in the sky, Andreae delivered a lengthy sermon on Christ's prophecy of the Last Judgement in Weimar. The audience probably expected that Andreae would talk about the comet, a topic suitable for a message on the world's end. Andreae did mention the comet in his sermon, but it was hardly the centre of his concern. He was preoccupied with one thing: how to promote the unity of Lutheran faith as based on the Formula of Concord, completed in May 1577. It had already received a warm endorsement from several Lutheran princes, including Elector August of Saxony. In his homeland of Württemberg, all pastors, teachers and Tübingen University professors had signed the Formula in July.[20] Andreae was filled with hope that the unity of Lutheran faith could soon be finally soon, and he used this Weimar sermon to explain the doctrines of the Formula and how they differ from the erroneous teachings of his opponents. The goal of this sermon was to teach the laity these doctrines in simple language so that they could defend their faith from the attacks of false teachers. The necessity of moral regeneration was discussed in the latter half of the sermon, and in this context Andreae mentioned the great comet.

The scriptural basis of this sermon is Christ's prophecy in Matthew 24, and Andreae explains that Christ teaches three different things in these passages. First, Christ warns of the destruction of Jerusalem and the fall of the Jewish Kingdom. Second, he explains what the condition of Christians prior to the Last Judgement will be. Third, Christ points to the consolation true believers can enjoy at the world's end.[21] In the subsequent pages, Andreae describes what the teachings of Christ mean for the German people of his age.

According to Andreae, Christ's warning and lamentation over the fate of the Jewish people shows the seriousness of the sin of contempt of God's Word. Germans have to take heed of this warning because this sin is also prevalent in Germany. One must seek God's Word earnestly and turn to God. Then Andreae turns to the history of the Church and examines how the devil, since the age of the Apostles, had been making his best efforts to destroy believers. First, the devil used tyrants like Roman emperors to crush true faith, but the Apostles resisted persecution with God's Word. Convinced that this tactic did not work, the devil now employed 'false teachers and Antichrists' to seduce believers. Soon the pope took power and introduced many human teachings and idolatry into the Church, but the power of the pope was broken by Luther. Then the devil devised splitting the Church between Anabaptists and Sacramentarians, and soon disunity spread even among those who adhered to the Augsburg Confession. This 'dangerous and heavy stumbling block' of endless disunity and theological struggles brings pain and doubt to many pious hearts, and such is the condition of Christians in the age of the Last Judgement. However, Andreae emphasizes that Christ promises true believers that they will be protected from idolatry and that Antichrist will finally be destroyed by God's Word.[22]

This is the end of the first part of the sermon, and it serves as an introduction to the second part: the examinations of controversial theological topics and exposition of the true, orthodox Lutheran view (as presented in the Formula). The exposition on controversial issues, such as original sin, free will, the meaning of good work and the nature of the Holy Supper, shows the essence of Andreae's educational method and pastoral care. He uses simple words that even the uneducated laymen can understand, and he emphasizes that one can judge the orthodoxy of doctrines based on materials familiar to ordinary laymen, especially the catechism. His arguments against dissenters are not necessarily always accurate. In his refutation of the Flacian stance of original sin, he deliberately avoids the use of key terms of the controversy, such as 'nature', 'substance' and 'accident', because he fears that such abstract concepts could confuse laymen. Consequently, his arguments oversimplify Flacius's theological understanding almost to the degree of 'malicious perversion', according to Kolb.[23] Andreae thunders that the Scripture tells us there are obvious differences between original sin and sinners, and one cannot equate a human being with original sin. The Ten Commandments forbid sinful acts of humankind, and are clearly based on the premise that sin and sinners should be distinguished. He even states that the Flacian stance is against the Lord's Prayer because

> we pray 'our Father'; but we cannot pray 'our original sin's Father' because not God but the devil is a father of original sin and all sins. Therefore there must be a difference between fallen children and original sin, otherwise we must learn a new Lord's Prayer.[24]

Such arguments totally miss the subtleness of Flacius's theology, and Flacius and Irenaeus would have felt insulted for being refuted in such a childish logic. However, it is Andreae's teaching method to instruct the laity to stick to simple principles and construct their arguments with the knowledge they had already received from the catechism.

After explaining ten controversial points, Andreae points out that even in this difficult time God raises Elector August of Saxony to promote the unity of the Church. It is God's grace to grant the unity of faith before the coming of Judgement. However, there is still danger brought on by the devil: the rise of the epicureans. Today people are filled with earthly worries as if they were created only for earthly lives. 'Isn't it clear that the Last Judgement is at hand', considering the condition of the world? At this moment Andreae quotes news of *Wunderzeichen* to enliven his argument. God showed a dreadful sign, namely 'a fiery rod' (a comet), like a faithful father who warns his children. However, Andreae also inserted a passage condemning astrology, declaring that we should interpret the meaning of the comet not by the cursed art of 'Chaldeans, Arabs, Egyptians and their disciples'. The curious world observes the movements of stars and predicts that such-and-such a prince will die or such-and-such a land, which is under the influence of the sign, will have misfortunes. However, such 'prodigy- and sign-readers' are not needed. Reminding hearers that Moses severely forbade such fortune telling, Andreae denounces astrology as 'spiritual fornication' and 'idolatry' that provokes God's wrath. Christian people should not have contact with such persons; otherwise they will be tainted. Indeed, one should not stay with such prophets of the devil, even for a moment. Since we already have the eternal truth through Christ, we do not need other 'sign-readers'.[25]

After these severe attacks on astrology, Andreae returns to the great comet again and explains 'how one should see it'. There are two kinds of people: pious, repentant people and evil, unbelieving people. For the former group, the celestial sign is 'a joyful sign of Grace' that foretells liberation from the sufferings in this world, eternal joy and salvation. However, for the latter group, God calls them to repentance and threatens them with his wrath through preaching of his Word and signs in heaven. Therefore, all pious Christians should stay in this interpretation given by Christ and flee from sign-readers as 'the devil's proclamation' ('Teuffels Bottschafft'). While such a devil's messenger says hundreds of lies, when his one prediction comes true, people put trust in him and they were taken away from God and his Word. When a Christian wants good weather, he should not rush to almanacs filled with lies; just turn to the Scripture and learn God's commandment. Here the words 'Deuteronomy 28' are written in the margin, and Andreae briefly returns to his arguments in the *Reminder*. After invoking the joy and consolation of the Second Coming of Christ, he concludes his sermon with a call to repentance and preparation for the approaching world's end.[26]

Though a large part is devoted to the explanation of the doctrines in the Formula, this sermon is a good example of Lutheran eschatological preaching, covering Christ's prophecy, the history of the Church which had always been under Satan's attack, the spread of vice among people, contempt of God's Word and the example of the Jews. *Wunderzeichen* are also not forgotten. Andreae cites the great comet to call people to repentance and moral regeneration, and he also notes that it means 'a sign of grace' ('gnadenzeichen') for those who repent while it means a sign of wrath for unbelievers. This twofold interpretation of celestial wonders often appears in *Wunderzeichen* discourses, and Andreae's use of the comet shows that he was familiar with these materials. Immediately after invoking news of the great comet, Andreae presents his condemnation of astrology in a much more sophisticated fashion than in the *Reminder*. The reason why he hated astrology becomes clear from his arguments: almanacs take away common people's attention to the Scripture, especially God's commandment, the true source of blessings; they make people depend more on astrologers' words than on God's Word. Probably he worried that if he introduced the story of a comet carelessly, it would arouse the people's interest in astrological predictions. Thus he chose to discredit astrology first, and only after he rejected astrology as the devil's trick, he presented the comet as a sign of God's will on each human being. Considering Andreae's deep concern with the laity, this may be why he does not show much interest in celestial phenomena in his writings.

Conclusion

Andreae's mistrust in astrology characterizes his attitude towards wonders in the sky. In 1568 he expressed his suspicion mainly because the accuracy of astrological calculations seemed questionable. By 1577 he no longer cared about the accuracy of predictions, but denounced the art as spiritual fornication and idolatry originating from the devil. Suspicion of astrology may have made him reluctant to discuss celestial phenomena extensively. We have seen that in the Weimar sermon he gives a Christian interpretation of the comet only after he denounces astrology. In the *Reminder*, he states that he does not need to discuss apocalyptic signs in the natural world because much has already been written. Most likely he thought that too much emphasis had been put on celestial wonders, and this cautious attitude shows a marked contrast with Irenaeus.

Nevertheless, Andreae was a practical man with rhetorical skills. He did not hesitate to employ *Wunderzeichen* stories and astrological imagery when he judged it was useful for his ministry. The image of five planets of vices controlling people's moral behaviour had to be appealing for people just impressed by unusual eclipses. He also used the concept of *Practick* and argued that the only true *prognosticon* one can rely on was God's Word, which promises blessing to those obedient to divine commandments. Andreae's twofold interpretation of the comet as 'the sign of Grace/wrath' indicates that he knew how to 'read' celestial phenomena in an orthodox Lutheran way and how to present it to the laity.

These facts show that what dominated Andreae's thinking throughout his career were his pastoral concerns. The sermons we examined show that he was making his best effort to instruct people and ensure their spiritual well-being based on the Scripture and the catechism. Though they were published in different decades (1568, 1577, 1588), the necessity of moral regeneration is a common thread in the three works. Close examination of his sermons reveals that his attack on epicurean living was not just conventional rhetoric of Lutheran jeremiads. His appeal to establish Christian discipline through collaboration between secular and ecclesiastical authorities suggests that he was seriously seeking a workable solution. However, there is one more important point: Andreae regarded the spread of epicurean lives in his days as part of Satan's conspiracy to destroy the Christian Church, placing it in the spiritual history of the Church. Since the beginning of Christianity, Satan had been attacking the Church through persecutors, the papacy, sects, false teachers and epicureans. As the doctrinal unity of Lutheran faith would be achieved soon, Andreae might have thought that the next chief threat was epicureans. It is no wonder that he quotes news of the great comet immediately after his arguments about epicureans, because the spread of sinful living in his age was in fulfilment of Christ's apocalyptic prophecy. The 'fiery rod' was a warning sent by God so that we do not show contempt for 'God's intimidation'.[27] However, even here Andreae does not forget his pastoral role and encourages the hearer with balanced, positive imagery. He compares God's showing of the comet with an affectionate father who is extending his hand towards his children.

This chapter shows how a conscientious pastor could guide listeners through his well-organized sermons. Andreae's effort to make his messages easier to understand and to accept for the laity shows the Lutheran preachers' ability to incorporate their parishioners' mental world into their messages. Even when they denounced people's pleasure-loving lifestyles, some of them could employ creative rhetoric and imagery familiar to the laity in order to draw their attention. Preachers did not simply indulge in dry, dogmatic arguments incomprehensible to the laity but rather encouraged parishioners to find how many things they could learn from the simple catechism, even to the degree that they could refute the errors of dissenters by themselves. Pastors like Andreae made conscious efforts to fill the intellectual gap between educated clergy and their parishioners, and celestial phenomena offered good material for the preachers to achieve this goal.

7 CELESTIAL WONDERS UNDER THE SHADOW OF WAR

In 1605 the Urbach minister Johann Sifard published a book titled *The Mirror of Comets* (*Cometen Spiegel*), which explored the history of comets and the misfortunes forewarned by these celestial phenomena. In the introduction of the book, Sifard states that 'it is natural that each sensible person loves [his] fatherland'. Our Holy Roman Empire is the 'forth and the last empire ['Monarchey']', as Daniel's prophecy shows, and people are now warned by God through celestial signs because of their ingratitude towards God's Word and *Sicherheit*. He declares that it is out of his love and sympathy for 'my fatherland Germany' that he published this small book to warn people.[1]

Though his statement can hardly be called unique, Sifard had enough reason to worry about the future of the Empire. The Formula of Concord finally established a confessional norm for the Lutheran Church, and it brought an end to decades of theological feuds among the clergy (though some theologians like Irenaeus continued their fight). However, the future of the Lutheran Church was still far from bright. Indeed, the years between Irenaeus's death and the outbreak of the Thirty Years' War was characterized by rapidly growing tensions between the confessional camps. Calvinists continued to expand their influence in Germany, while the restoration of Roman Catholicism was also in progress in many territories. It was clear that the mentally troubled emperor, Rudolf II, was losing control over the confessional power balance in his Empire. Fearing that a military clash with the Catholic powers would become inevitable, some Protestant states adopted a more confrontational policy towards the Habsburgs, though the reactions of the Protestant princes were far from united. Moreover, the Lutheran Church was facing various new challenges from within. Proto-pietism, well summarized by Johann Arndt's book *True Christianity*, severely criticized the current condition of Lutheran religiosity and called for spiritual renewal. Various occult faiths and esoteric ideas that deviated from orthodox Lutheran teachings spread among intellectuals, and the writings of the so-called Rosicrucians, who first appeared in the 1610s, were welcomed enthusiastically by those longing for a 'new reformation' of the world.

Faced with such growing tensions and confusions, it is no wonder that people maintained their interest in *Wunderzeichen* and were eager to find meanings in these phenomena. Prophetic literature, often coloured by mysterious language and images, appeared in this pre-war period and stimulated people's imagination. German people's 'singular' enthusiasm for prophecy and celestial phenomena surprised many foreign visitors; as the English traveller Fynes Moryson records in his famous *Itinerary* (1617), 'the Germans not only of the Comon sort, but of them that are not unlearned, give too much Creditt to predictions (which they call Prophecies) of their owne Countrymen ... In generall the Germans seeme to have singular credulity towards forraine Prophetts, and Astronomers, but especially to their owne'.[2]

Did such changes in politics influence the contents of discourses on celestial wonders? Was there any relationship between new trends in Lutheran spirituality and the *Wunderzeichen* literature during this period? These questions are the topic of the following sections. First, in order to understand the context of *Wunderzeichen* discourses during this period, we shall briefly review the political and confessional conditions surrounding the Lutheran Church between 1595 and 1618. Then we shall examine the general tendencies of *Wunderzeichen* discourses in the same period and analyse whether there was any change in the basic tenets of *Wunderzeichen* messages. This analysis is followed by an examination of how Arndt and the anonymous author of the Rosicrucian manifestos incorporated stories of wonder into their own writings.

Before entering the phase of the Thirty Years' War, we should also consider the limitations of the spiritual scope of *Wunderzeichen* discourses, which increasingly became clear as Lutheran consolation literature became popular in the late sixteenth and early seventeenth centuries. The weakness of *Wunderzeichen* discourses became manifest when we compare them with other kind of wonder stories, namely the sermons on climatic disasters and the reports of encounters with angels. We shall see what these two kinds of literature provided readers with what *Wunderzeichen* works could not. Finally, we shall read several sources on the great comet of 1618 and consider why the production of *Wunderzeichen* literature declined during and after the war. Through these examinations, it becomes clear that *Wunderzeichen* discourses kept their popularity because the Lutheran eschatological world view, the theological backbone of literature on wonders, remained unshaken until the latter half of the Thirty Years' War.

The Political and Religious Conditions of the Empire, 1595–1618

Between 1595 and 1605 a number of important Lutheran territories, such as Saxony-Anhalt (1596), the Upper Palatine (1595–1598) and Hesse-Kassel (1605), went over to the Calvinist camp. In 1613 Margrave Johann Sigismund of Brandenburg made his conversion to Calvinism public and tried to impose the Reformed faith upon his subjects. He had to abandon his plan due to the widespread protests of the pastors and his subjects,[3] but Sigismund's case showed that despite the Lutheran clergy making careful efforts to prevent it, Calvinist influences crept into princes' hearts.

The steady advance of the Counter-Reformation was also a serious concern for the Lutherans. Although the Wars of Religion in France finally came to an end (with the conversion of Henri IV to Catholicism), the revolt in the Netherlands was still continuing. In the Empire, the ecclesiastical princes of Fulda, Würzburg and Bamberg gradually restored Catholic hegemony in their territories through the introduction of the Tridentine Reform and the expulsion of their Protestant subjects. In 1607 Duke Maximilian I of Bavaria sent an army to Donauwörth and restored Catholic worship by force, although he had no direct, legitimate authority over the city. This incident clearly showed that Catholic forces and Protestant princes were on a collision course. Sporadic persecutions of Protestants also broke out in Habsburg Austria, as shown by the case of the expulsion of the leading Protestants of Graz in 1600, ordered by Archduke Ferdinand.[4] Although Emperor Rudolf II was hardly enthusiastic about the papal programmes of the Counter-Reformation, he was suffering from a decline of mental health, and he constantly changed his religious policy. The mental condition of the emperor was a serious political issue because if the Protestant electors declared heirless Rudolf incompetent to rule, they could seize power and prepare a way to elect a non-Habsburg emperor. Before this occurred, Rudolf's brother Matthias took away the power from the emperor by force and later became emperor himself (in 1612). However, since both Matthias and Rudolf lacked an heir, it was simply a matter of time before a succession crisis would break out.

As tensions between Catholics and Protestants rapidly grew, some Protestant states felt they should prepare for the worst-case scenario. The Calvinist leaders of Electoral Palatine expected a major religious war to be unavoidable and believed that the formation of a military alliance of Protestant princes was an urgent task. Shocked by the Bavarian occupation of Donauwörth, several Protestant princes and cities formed the Protestant Union in 1608 to defend themselves from the aggression of papist forces. However, some major Lutheran states such as Electoral Saxony refused to collaborate with the Calvinist princes and did not participate in this alliance. The grave concern about this disunity among Protestant princes was expressed by Landgrave Moritz of Hesse in 1615:

> I am very much afraid that the states of the Empire, quarrelling fiercely among themselves, may start a fatal conflagration embracing not only themselves ... but also all those countries that are in one way or another connected with Germany. All this will undoubtedly produce the most dangerous consequences, bringing about the total collapse and unavoidable alteration in the present state of Germany.[5]

Sensing approaching danger, many territories built or repaired fortifications of cities to prepare for an outbreak of war.[6] Thus a tense atmosphere prevailed in the Empire during the period immediately preceding the outbreak of the Thirty Years' War.

Wunderzeichen Discourses, 1595–1618: The General Picture

Though a great number of broadsheets, tracts, books and sermons concerning celestial wonders were published during this period, if we compare them with the *Wunderzeichen* discourses examined in Chapters 1 and 2, it is difficult to find any substantial change in the content of these messages. Christ's prophecy of signs of the End Times is regularly invoked, and they end with the familiar conclusion: repent or disaster!

In November 1605 *Chasmata* (fiery signs) were observed in the sky over Darmstadt throughout the night, and Heinrich Leuchter, the court preacher, delivered a sermon on this phenomenon. Leuchter reminds readers that Christ said that signs would appear prior to the Last Judgement. Or, he adds, this is the sign of great misfortune and punishment we had earned by our many sins, but God, through his grace, showed this sign to us as a warning. Quoting the story of the celestial vision of an army that had appeared in Jerusalem (in Maccabees), he explains that God took 'exhalation [from the earth] or materials from nature, namely mist and damp', and made amazing visions from these elements. All kinds of fiery signs we often see in the sky are created in this way. Then Leuchter describes some stories of *Wunderzeichen*, including the great comet of 1577 and the new 'wonder-star' of 1604, and states that such signs generally signify 'something bad'. The familiar lamentation over unfaithful Germany is repeated: though the Gospel has been preached for 100 years, we keep returning to so many evils. 'The longer it stands, the more a young tree bears better fruit. Yet in our case, the longer we live, the more we become outrageous'; therefore 'let us pray, let us repent and turn from sins. Let us be pious, praise and fear the Lord'.[7] Leuchter's view that God creates celestial visions by using mist and damp rising from the earth, obviously taken from the theory of comet formation, may be unique, but otherwise his message contains no new elements.

Some preachers pondered possible connections between celestial phenomena and the tense political condition of their day, although they did not pursue it too deeply. One example is *A Good-hearted Warning Writing about Future Misfortune on our Beloved Fatherland German Nation* (*Eine guthertzige warnungsschrifft für künfftigem Vngluck vnsers lieben Vaterlandes Deutscher Nation*; 1609), written by Stephan Hering, pastor of Gottleuben in Saxony. As we can deduce from the title, Hering's main message is that sinful Germany is facing the grave danger of God's punishment. The author also quotes stories of *Wunder-*

zeichen to press readers to amend their lives. He reminds them of the *Chasma* of September 1604, which '[did] not differ from an artist's painting of the fiery pool of sulphur over Sodom and Gomorrah', and states that it can be connected with Christ's prophecy that 'the heaven will be shaken'. Indeed, there are so many signs shown in the sky by 'patient God'. There were fifty-one eclipses of the sun and the moon between 1590 and 1607, plus three additional ones observed in 1608.[8] 'Comets or wonder-stars' should not be forgotten because Christ said that signs would appear in the stars. 'Experience tell us that God always shows wonder-stars when he announces his wrath and will inflict punishment'.

After narrating more examples of celestial phenomena, Hering quotes historical examples of a fiery sign seen before warfare. The latest example he mentions is the one that appeared in December 1604, 'that from the beginning of the world, such one had not been seen, as *Astrologi* admit'. Hering states that this sign appeared during the time of the 'defeat' and massacres in Hungary, 'the bloodbath of the devil [worked] through his instrument the Jesuits'.[9] The lesson of these supernatural signs is, of course, to repent and abandon sinful lives: 'the door of God's Grace has not been closed yet'. The bloodbath mentioned here must be the persecution of the Protestants in Hungary and Transylvania during the Long Turkish War (1593–1606). Emperor Rudolf stationed his troops in the region to stop the advance of the Turks, but he used the presence of the troops to restore Roman Catholic worship, though there were only few Catholics in these territories.[10] The spread of persecutions from Austrian territories to Hungary must have been shocking to the Lutheran clergy, and it was no wonder that they suspected that the Jesuits were behind such a 'bloodbath of the devil'. This reflects Hering's serious concern about the increasingly dangerous situation faced by the Protestants.

The next year, the people of Hesse experienced a number of wondrous phenomena. A comet was seen in March, followed by three suns observed in the sky the next day; earthquakes and storms surprised people, and a terrible drought caused the deaths of many domestic animals. But the strangest phenomenon of the year allegedly occurred on 30 July: an apparition of a flying dragon with dreadful fiery eyes appeared in the evening sky, followed by *Chasmata* and thunderous weather. Hartmann Braun, pastor of Grünberg in Darmstadt-Hesse and one of the most prolific authors of printed sermons in his day, delivered a sermon on this mysterious dragon, published in Darmstadt the next year. In the introduction, Braun states that the sequence of prodigious incidents of 1615 was not just a warning of great misfortune or of punishment but also a sign of the *dies vltionis*, the Day of the End. After invoking Christ's warning that no one knows when the Day of the Lord will come (Mark 13) and admonishing hearers to always be alert, Braun turns his argument to the dragon. The text that he chose for his exposition is the section about the apocalyptic dragon in the twelfth chapter of the Book of Revelation. The monster is called the 'red dragon' because it is a 'bloodthirsty helldog' that causes war, murder and persecutions of Christians. The fact that it has seven heads indicates that the monster signifies Rome because the city is founded on seven hills.[11] Braun points out that a dragon can also signify tyrants and persecutors of the Church, such as Herod, Nero, the Turks and the pope.

Here we may expect that Braun will express his opinion about why a dragon, the symbol of Antichristian Rome and persecutor of the true Church, appeared in the sky at this time. However, Braun does not direct his argument to contemporary political situations. While recounting various historical visions of dragons, he mentions only one case related to the persecution of Protestants: in 1554 a fiery dragon was seen in the sky over France, and it was followed by the burnings of many for religious reasons. However, regarding the dragon in 1615, he simply states, 'what this and such wonder and signs mean and suggest, only God knows', and urges people to repent and pray to God. 'Our dear Father, turn away the implied future misfortune and deserved punishment from our dear Fatherland. Protect your small herds of righteous Christians. Under princes, maintain peace for the subjects, and save them from rebellion, war, and bloodshed'.[12] The tone of Braun's prayer suggests his concern that this dragon was the sign of coming war and persecution, but politics are not mentioned in the sermon; his message is kept in the framework of a conventional call to repentance.[13]

The above are only a handful of examples published during this period, but they show the general characteristics of *Wunderzeichen* discourses at this time. The clergy undoubtedly retained their enthusiasm for celestial phenomena, and sometimes showed an extensive knowledge of *Wunderzeichen* news in their writings. The accumulation of *Wunderzeichen* literature over the previous fifty years offered a wealthy resource for preachers who wanted to use stories of wonder in their messages. Some of the writers also showed interest in recent astronomical studies and used them as references in their apocalyptic discourses.[14] The strained atmosphere of pre-war society had to stimulate people's interest in strange phenomena, and in some of these discourses we can see the authors' concern about religious persecution and unrest. However, these authors generally avoided direct discussion of contemporary politics. As Lutheran preachers, they knew that their primary task was to call for repentance, and they believed that it was the best defence from God's wrath and disasters.

New Trends in Lutheran Spirituality and *Wunderzeichen* Discourses

During this pre-war period the Lutheran Church received serious challenges from within, especially brought about by two new trends of spirituality: proto-pietism, represented by Johann Arndt's devotional theology, and occult faiths, which mixed Lutheran theology and esoteric magical thoughts. Was there any relation between these trends and contemporary discourses on celestial phenomena? Did their emergence affect the contents of the clergy's *Wunderzeichen* discourses?

The history of German pietism cannot be recounted without reference to the name of Johann Arndt, the pastor of Braunschweig who later became the General Superintendent of Celle. In 1605 he published the first part of his most influential work, *True Christianity*, and the complete volume, consisting of six books, appeared in 1610.[15] Arndt's thought can be summarized in one passage from the volume: 'If we are to become new creatures by faith, we must live in accordance with the new birth ... It is not enough to know God's word; one must also practice it in a living, active manner'.[16] Arndt does not hide his aversion to the theological polemics of his age, which caused more conflicts than love. 'It is not enough only to write against sects and heretics, to preach and dispute, to maintain pure doctrine and true religious'. He laments that

> [t]hese [polemical] activities have fallen into great misuse in our time ... It is as if Christianity consisted only in disputations and the production of polemical books, and not far more in seeing to it that the Holy Gospel and the teaching of Christ is practiced in a holy life.[17]

A new wind, completely different from the fierce cries of Irenaeus, is blowing here.

How did Arndt interpret celestial signs in his age? In the fourth book of *True Christianity*, he focuses on the world of nature based on the story of creation in Genesis, and praises the universe as a magnificent expression of God's glory and love. God assigns a role to the heavenly bodies: to show 'signs of time'. They foretell the second coming of Christ and also give warnings of God's punishment. 'Here we must consider, that God does not punish of a sudden, but does first warn us by Signs and Prodigies of the approaching Evils'. Arndt later returns to this topic and focuses especially on eclipses, '[t]okens of the Divine Wrath, Prognisticks of great Calamities and Revolutions in the World'. He believes that even if it occurs naturally, an eclipse is essentially an unnatural phenomenon. Since the sun and the moon were created to give light to the earth, if this natural role is hindered, 'they are then in an unnatural State'. The sun and the moon are 'the Mirrors of the great World' that reflect sins and iniquities among human beings. An eclipse 'points out to us that internal and spiritual Blindness of Heart, that reigns in every one of us' and it can be called 'a kind of penitential Sermons, preached by God to an impenitent World'. He also insists that the same can be said about fiery signs in the sky, which 'seems to speak to us in Words like these: Look up to me, and think of that Day when I shall burn with real Flames'.[18]

True Christianity is not the only source that shows Arndt's opinion on celestial matters. In *Postilla* (1616), his exposition on Christ's prophecy of the world's end, Arndt shows his extensive knowledge of the history of *Wunderzeichen*. He argues that the celestial signs of the end appear clearly and frequently so that no one can excuse himself by claiming that he did not receive the warning. Eclipses and other signs in the sun signify the condition of earthly *Regiment* and religion, and several historical examples can be quoted. The oldest one Arndt narrates is the three suns that appeared when Julius Caesar was murdered. The three suns also appeared when Vienna was besieged by the Turks in 1529, and the same phenomenon was observed in 1541 and 1549, signifying the Schmalkaldic War. In December 1602 a great eclipse occurred in the sun, and he states that this could be compared to the present condition of the world and religion: there was a disputation and colloquy between the Catholics and the Lutherans at Regensburg in November 1612. 'May God sustain the *Regiment* and the true religion', Arndt adds. An eclipse of the sun was also observed in May 1612. While eclipses and other signs in the moon can spell trouble in government, authority or religion, it can also mean troubles among subjects, such as 'revolt, rebellion, lies and betrayal'. These signs, foretold by the prophets, the Apostles and Christ himself, also proclaim the coming of the Last Judgement, 'not only to the world in general but especially to each individual'.[19]

The above writings show that Arndt was not indifferent to the *Wunderzeichen* of his days and sometimes connected them to contemporary events. However, though his reference to God's Judgement on individuals may reflect his concern with the significance of the individual's inner piety, Arndt's view on celestial signs does not show much difference from his predecessors' views. Here we have only examined Arndt's writings, but we can safely state that in general proto-pietism did not affect the basic tenets of *Wunderzeichen* discourses mainly because the new spiritual trend maintained the traditional framework of Lutheran apocalypticism. This supports Robin Barnes's view that '[m]ystical, individualizing tendencies had grown in Lutheran eschatology in the last decades of the sixteenth century and the first years of the seventeenth, but during that era these tendencies had remained within the frame of the apocalyptic world-picture'.[20] Irenaeus and other theologians regarded the prevalence of false teachings as a sign of the End Times and made their best efforts to crush these errors with polemics. But Arndt found that many of these passionate servants of Christ were running on the wrong track. It is no wonder that Arndt cries in his postil, 'it is clear that the world is hastily sinking, and the Last Judgement is near', and it is natural that he showed interest in celestial phenomena.[21]

The increasing popularity of occult faith also troubled the Lutheran Church in early seventeenth-century Germany. Since Barnes discusses this topic extensively in his study, I want to limit the scope of my analysis to the relationship between *Wunderzeichen* and one of the most famous examples of esoteric literature in this age: the Rosicrucian manifestos. In 1614 and 1615 two anonymous tracts titled *Fama Fraternitatis* and *Confessio Fraternitatis* were published in

Kassel and sparked the famous Rosicrucian furore throughout the Empire. The message of these tracts can be summarized as follows: a young German nobleman named C. R. (Christian Rosencreutz) went to the East for pilgrimage, and there he learned about the true ancient wisdom of universal harmony. After he returned to Europe, C. R. created a secret fraternity with several wise men to preserve this knowledge. The fraternity remained secret for one hundred years, but recently the grave of C. R. was discovered, and this signalled the beginning of a new era. The author calls those who want to join the fraternity to step up:

> We ought therefore here to observe well, and make it known unto everyone, that God hath certainly and most assuredly concluded to send and grant to the world before her end, which presently thereupon shall ensue, such a truth, light, life, and glory, as the first man Adam had.[22]

From the two tracts it can be deduced that the year of the discovery of C. R.'s tomb is 1604. Indeed, the year 1604 is suitable for this portentous incident. The *Confessio* states that

> the Lord God hath already sent before certain messengers, which should testify his will, to wit, some new stars, which do appear and are seen in the firmament in *Serpentario* and *Cygno*, which signify and give themselves known to everyone, that they are powerful *Signacula* of great weighty matters.

The signs mentioned here refer to the little nova in Serpentario (1603/4) and the comet in the sign of Cygnus.[23] Though such celestial phenomena were usually regarded as signs of the world's end, here they are depicted as signs of a 'new Reformation', a new golden age to come.

This message sounds like a new trend in *Wunderzeichen* discourses, but in reality the message of these Rosicrucian manifestos still remains within the framework of Lutheran eschatology. They do not deny the approaching end of the world, and in fact, as the above quotation shows, the coming of the end is the very premise of the 'new reformation' and restoration of truth. *Confessio* foretells the downfall of Antichrist (the pope), the familiar topic of apocalyptic prophecy, which 'is delayed, and kept for our times'.[24] Thus traditional Lutheran eschatology and the chiliastic Rosicrucian world view share the same root. Actually, the notion of the new age foretold by *Wunderzeichen* was neither new nor original. In Chapter 2 we saw that many clergy regarded the 1572 nova as a sign of the imminent Second Coming. Though the Rosicrucian interpretation of the 1604 nova seems to put more weight on the reformation/transformation of this world, such a change was regarded as an apocalyptic event that would soon be followed by the real end of the world. In addition, the expectation of a new golden age prior to the end was not rare in Germany throughout the sixteenth century, along with predictions of the Last Judgement. As we saw in Chapter 6, in his astrological tract on the 1567 eclipse, Paul Severus predicts that after various tribulations 'a new Reformation' will take place in 1570 and a pious emperor will restore Christianity in Hungary and Constantinople. Such a prophecy of a golden age ruled by a messianic emperor had been popular since the Middle Ages.

The author of *Fama* and *Confessio* knew the appealing power of *Wunderzeichen* well. By incorporating the story of the nova, he tried to strengthen his argument that a new age had started. Indeed, it seems that his strategy worked. In 1618 a 'Theophilius Schweighardt' published a book that contains an illustration of the 'invisible college of the Rosicrucians', and in this engraving the 1604 nova and the comet are shining in the sky above the Rosicrucian building.[25] This shows that the connection between the Rosicrucians and celestial phenomena left a strong impression on people's imagination. Several historians discuss the possible influence of the writings of Simon Studion, an obscure pastor of Marbach near Stuttgart, on these Rosicrucian writings. Based on complicated calculations inspired by the books of the prophets in the Old Testament, Studion predicted that the pope would be crucified around 1620 and the coming of Christ's Kingdom Second would take place in 1623.[26] In one manuscript, Studion regards the 1572 nova as the key to calculate the timetable of the end. Probably the author of the Rosicrucian manifesto recalled Studion's chiliastic idea of the new star when he saw the nova in 1604 and incorporated it in his writings.

Who was the Rosicrucian author, who cleverly exploited the popularity of celestial wonders? Many scholars agree that these manifestos were produced within the circle of Johann Valentin Andreae – none other than the grandson of Jacob Andreae. J. V. Andreae, who later became the Superintendent of Calw, probably did not write the manifestos himself, but there is a strong possibility that he was involved in the production and circulation of these tracts.[27] Indeed, he seems to know the work of Studion because he mentions it in one of his books. In his famous utopian fiction *Christianopolis* (1619), J. V. Andreae inserts one passage that praises God's grace revealed through the heavens:

> The sun, the stars, the rainbow, hailstorms, the dew – to name but a few – what wonderful benefits they confer on men of faith! ... It teaches by means of portents and miracles. It thunders against irreligion. It raises the heads of godly men and makes them look upwards towards the hope of restoration.[28]

This passage can be interpreted as an expression of the author's traditional Lutheran view on celestial matters (people see the signs of God's wrath against disbelief and the coming of the Last Judgement, the time of ultimate 'restoration', in the sky), but it also reminds us of the message of the Rosicrucian manifestos. The Rosicrucians looked up to heaven and hoped for 'restoration' of the original truth in the 1604 nova and comet, but their thought was still firmly founded on traditional Lutheran eschatology.

Limitations of the *Wunderzeichen* Message in Comparison with the Discourses on Climatic Disasters and Angels

Although there was no outstanding change in the contents of *Wunderzeichen* discourses, when we compare *Wunderzeichen* literature with publications on other kinds of wondrous phenomena, we get the impression that the limitations of the spiritual scope of the *Wunderzeichen* message became clear during the late sixteenth and early seventeenth centuries. The sources examined in this section are religious discourses on climatic disasters and encounters with angels. Both climatic catastrophes and appearances of spirits were regarded as wondrous signs sent by God, and various reports of these incidents were incorporated in wonder books such as Fincel's. However, the texts that discuss these topics contain several points which are not covered in *Wunderzeichen* literature.

As the climatic condition of Central Europe deteriorated after 1560, the Lutheran clergy had to respond to their parishioners' concerns about why they had to suffer from increasing storms and other climatic anomalies, and what they should do to escape harvest failures. Their responses created a genre of sermons called *Wetterpredigten* (sermons on storms). The structures and messages of these sermons show a remarkable degree of consistency, basically covering the same four points: where storms come from – God is the origin of all climatic phenomena; why God brings storms to people – to punish human beings for their sins and to warn them of the imminent coming of the Last Judgement; how people should respond to such disasters – by repenting and amending their lives; and what kind of consolation Christians can have during storms – believers can put their trust in God even in the midst of storms.[29] Though the similarity between the message of *Wetterpredigten* and that of *Wunderzeichen* literature is obvious, there are several differences between the two genres of discourse. First of all, while *Wunderzeichen* prints were essentially 'preaching of the Law' and focused on God's wrath on human sins, storm sermons often emphasized God's love and mercy, which were promised to repentant believers, and encouraged readers not to fall in despair but put trust in his providence.

One early example is David Bramer's *On Thunder, Lightening, Hailstorm, Stormy Wind, and Other Kind of Great Storms* (*Vom Donner Blitz Hagel Sturmwinden vnd andern grossen Vngewittern*; 1577), the work that could be called the pioneer of *Wetterpredigten* literature. In this book, Bramer, the Superintendent of Saalfeld, stresses that storms are God's sermons calling people to repentance; but in the final part of the book, he encourages readers to consider that storms are coming from God, the loving Father who can protect his children from any harm. God can restore the properties the believers had lost due to climatic disasters, as he did for Job after the test of faith. God causes 'all things to work together for good to those who love God', as Paul states in Romans 8. The loss of family and friends is irrecoverable, but one should remember that none in this world is free from the danger of death. We should flee to God's hands, keeping Paul's words in Romans 14 in our minds: 'whether we live or die, we are the Lord's'.[30] This same passage of Romans 14 is often quoted throughout the *Wetterpredigten* of the early seventeenth century to call people to put their hope in God and to trust his will.[31] The authors' delicate pastoral consideration to provide people spiritual encouragement is clear in these texts.

Although such consolatory emphasis on God's caring love is not absent in sixteenth-century *Wunderzeichen* discourses, as we saw in Pfeffinger's work in Chapter 2 (Pfeffinger also quotes Romans 14), it is hardly the main point of these texts. This difference can be explained by the fact that *Wetterpredigten* were delivered to encourage the victims of disasters, while *Wunderzeichen* prints were intended to threaten those living in *Sicherheit* with imminent punishment. However, as Soergel points out, this growing emphasis on God's love may indicate the gradual shift of Lutheran spirituality in the early seventeenth century. According to Soergel, the increase of consolation sermons on local disasters reveals 'a new willingness on the part of the Lutheran clergy to engage with the psychological effects of great tragedies rather than to merely warn about their moral and eschatological significance'.[32]

Reflecting this trend, some early seventeenth-century *Wunderzeichen* works placed more emphasis on God's care and tenderness than the works in the previous century had done. In his 1605 comet book, Sifard states that God is so faithful that he gives people time for repentance, as it is said: 'Good friend is the one who warns others'. He assures that God is 'not our enemy, wrathful judge and punisher, but far more our friend' who seeks our best.[33] However, the more outstanding example is the sermon delivered by Laucha pastor Johann Mülmann on the fiery celestial sign in December 1604. In the sermon, Mülmann regards the phenomenon as a clear sign of the imminent coming of Judgement Day and urges people to repent, but he also does not forget to remind them of God's tender love with various lyrical metaphors. God sets two large candles, namely God's Word and celestial signs (such as the one seen in December), to show those in the darkness of sins the right way to heaven. Even in his wrath, God's love and mercy cannot be hidden. God 'gets angry with us, but he does not hate us because his wrath is directed for our improvement'. God says, 'I have looked for you, as a shepherd looked for his lost sheep. I have searched for you, as a woman searched for her lost coin [Luke 15]'.[34] In the conclusion, Mülmann admonishes people to lighten the lamps in their hearts with the oil of faith to wait for Christ's return. People should not become timid because of these fiery signs; rather, they should raise their heads because their salvation is near. Then the 'firework of God' (the fiery sign) is nothing but a torch of the bride of Christ placed in heaven, and in the evening of this world, Christ will lead the elect to the heavenly fatherland with it. Though the content of the main message is the same, the difference of its tone from traditional *Wunderzeichen* discourses is clear. These examples may suggest that the conventional discourses on celestial phenomena, which focus only on the strict Law and the image of the wrathful God, had a serious limitation when they were used for pastoral care.

Another genre of literature to compare with *Wunderzeichen* literature is the reports of encounters with angels. Though there are a variety of stories on this topic, many reports claim that a mysterious figure (sometimes more than one) dressed in white suddenly appeared to a person and entrusted a message to the local community, usually a warning of imminent divine punishment. The main message of these stories is the same as for *Wunderzeichen* news, and such reports were not rare in Lutheran Germany and Scandinavia during the sixteenth and seventeenth centuries. For example, in 1595 Jacob Coler published a tract on the strange case of a twelve-year-old girl in Berlin who claimed that a supernatural figure in white with a sword in his hand suddenly appeared to her during the night. The angelic man conveyed to her that Germany would be punished soon because of people's arrogance, represented by their fashions. The rumour of the girl's encounter immediately spread among the populace in Berlin. Coler was at first sceptical, but as he summoned the girl and investigated the matter, he was convinced that this manifestation was a divine warning not only for Brandenberg but also for all of Germany. An angel was sent to the earth at the times of Christ's birth, death and resurrection. If so, Coler asks in his tract, why not at the time of his return?[35]

Some records indicate that people showed enthusiastic responses to the news of the manifestation of good angels. In the Berlin case, Coler launched his investigation because the rumours of the angelic encounter became rampant and started spreading to other regions. What made these stories of angels so appealing to the populace? The case of Anna Schützen of Dürrmenz in Württemberg offers us a clue. In May 1563 a mysterious young man in white suddenly appeared to Anna, wife of a day labourer, and told her that God was angry with the wealthy who did not care for the poor. After this initial encounter, the angelic man appeared to her several times, and as a sign of the truth of her experience, the angel made her sit for hours on a bench, unmovable. The news spread immediately, and upon hearing the rumour that Anna would speak in public, about 1,500 people came to her village. The excitement of the crowd is well attested by the fact that some people who had talked with Anna 'jumped around with joy and thanked God for having seen the angel woman and for having spoken to her. They would not exchange this for the best meal, but a woman said she did not view God as dearly as her'.[36]

Such reactions may be extreme, but it shows the common people's intense desire to touch the supernatural directly. Stories of angelic encounters were appealing because they assured people that God still cared for them (especially the poor) and intervened in human affairs in direct ways (by sending angels). Indeed, the presence of angels on the earth itself shows God's will to protect and guide his people. In his pamphlet on the flood that struck the town of Öhringen in 1589, David Meder, the Superintendent of the city, points out that during the flood a mysterious fiery light was seen near the city wall. Meder's interpretation that the light was an angel sent by God to help the poor suggests his desire to find a sign of God's care even in the midst of disaster.[37] Even if the angel's message is a gloomy warning of divine punishment, the possibility of direct physical contact between the divine and human beings through angels had to be encouraging for people.

Wunderzeichen discourses do not have this appeal. Signs appeared in remote heaven, and people were expected to see them in awe, but there was no possibility of personal interaction with the divine world. Unlike the clear verbal messages of angels, interpretation of the meanings of the celestial signs was trusted to intellectuals such as the clergy and astrologers. This shows another limitation of the *Wunderzeichen* messages. We can observe this difference between the stories of celestial wonders and those of angelic encounters in Johann Schütz's brief tract on the celestial apparition that appeared in Bitterfeld in Saxony in 1568, which is mentioned briefly in Chapter 1. This report is a rare combination of a celestial phenomenon and a manifestation of an angel. A maidservant went to a women's garden (*Frawen Garten*) to bleach cloth, and there she saw a lovely boy dressed in white standing under a tree. He invited her to come near him, and as she obeyed, she found that the boy, who looked two years old, had small wings on his back. He told her not to fear and stated that he was a messenger of God who came to urge the authorities and clergy to call people to repentance. He then explained that the recent fall of blood from the sky and the celestial apparition of the blood-red sword meant bloodshed to come if people did not repent. Dreadful diseases and inflation will also befall the world. When the angel said that he had to go now, the maid wept bitterly and wiped her eyes with her apron.[38] Though the contents of the angel boy's message are as gloomy as other *Wunderzeichen* discourses, a warm, fairy-tale-like atmosphere and a personal touch underlie the story. The maid wept as the angelic child left, but we receive the impression that she was sadder about his leaving than about the dreadful contents of his message. We do not know whether Schütz, who wrote several tracts on wondrous incidents, invented the story himself or composed it based on what he heard from others; but by combining the story of celestial phenomena with that of an angelic encounter, he composed a story with a largely different atmosphere from other *Wunderzeichen* works.

Thus the comparison of *Wunderzeichen* literature with *Wetterpredigten* and angel stories reveals the fact that reports of celestial wonders alone could not satisfy the people's desire for spiritual consolation in the challenging situations of their times. While these limitations gradually became clear, *Wunderzeichen* literature entered the stormy period of the Thirty Years' War.

The 1618 Comet and the Outbreak of the Thirty Years' War

In early 1619 a Laingen broadsheet exclaims a familiar phrase: 'Wake up, wake up Christian people! In this dangerous time, [wake up] from the sleep of sin, and also from misfortunes'. In November 1618 a comet with a long tail had appeared from the East and flew to the West. Since comets never carried good meanings, the author states, this one also foretells bad news: inflation, war and pestilence will come. Therefore, the author urges, we must fall at God's feet and repent with sincere contrition so that he will not doom us. The broadsheet also recounts the news that an angel appeared in the city of Augsburg in January, and mysterious voices were heard saying, 'O men, amend your way' and 'alas, alas for this city'. After the angel vanished, two other angels in white also appeared. The author of the text uses this opportunity to attack the Jesuits in Augsburg. For him, this beautiful city was infected by 'many false teachings, namely the herds of the Jesuits, who fight against God and mislead many people every day'.[39] We have heard the same warnings of impending misfortunes countless times, but this time the matter was different. Real disasters finally came to Germany – the Thirty Years' War broke out with the Protestant rebellion in Bohemia. It is suggestive that the author of the Laingen broadsheet denounces the Jesuits in this text, anticipating the outbreak of war.

The 1618 comet caused an unprecedented flood of *Wunderzeichen* writings. According to John Theibault, more than one hundred pamphlets about this comet were published in Germany.[40] Many pastors discussed the meaning of the comet in their sermons on the second Sunday of Advent of the year, and some of them were published immediately. Most of these sermons were highly conventional 'repent or disaster' messages, and here we examine one example.

In his *Ulm's Comet Sermon* (*Ulmische Coemten Predigte*; 1619), Conrad Dietrich, the Superintendent of Ulm, thunders that since this *sichere Welt* does not listen to our verbal sermons, God now sent a new comet as 'another kind of preacher'. With 'his long light-emanating fiery rod', God preached to the world his wrath and future misfortune from the sky. In the midst of the sermon, Dietrich quotes various historical examples of comets and narrates what kinds of misfortunes followed; he states that the rise of *Rotten*, sects, heresy and splits in religious matters often follow the appearance of comets. For example, he recounts that the comet of 1572 (nova) and 1577–8 foretold the 'dreadful persecution of the Evangelicals in France, the Netherlands, and Cologne'. The 1580 comet signified the disputes over the Book of Concord, and the 1582 comet warned of the introduction of the Gregorian Calendar. The persecution of Protestants in Steyrmarck was signified by the 1596 comet, while the 1604 comet foretold the *Calvinische Reformation* in Marburg.[41] For Dietrich, all of these events of confessional struggle had cosmological significance.

Of course, he was not indifferent to the political condition of Bohemia. After quoting Luther's lamentation over people's *Sicherheit*, Dietrich expresses his concern about the careless (*sorglich*) atmosphere of German people and reminds them that the 'dangerous fires in Bohemia and Austria' that 'sparkle' could also fly to Germany. But he immediately adds his concern for the well-being of the emperor: 'God protect the pious emperor, the head in this cancerous time. Alas, German nation. What will you do after [this] strange play?'[42] By this passage, Dietrich reveals his conservative Lutheran attitude towards the succession problem in Bohemia. Until the new emperor's aggressive Counter-Reformation policy became clear in the late 1620s, many Lutheran preachers showed a marked reluctance to support the Protestant rebellion against the Catholic emperor, their legitimate ruler. The particular fact that the new king of Bohemia, Friedrich V of Palatine, was a Calvinist made their attitude towards the rebellion cool.[43] This gives us a clue as to why Lutheran clerics such as Hering and Braun were reluctant to refer to the political condition of the Empire in their *Wunderzeichen* messages. They obviously saw connections between the phenomena they observed and contemporary events, and they feared the possibility of the outbreak of war and persecution. However, they could not criticize the emperor and the princes directly. Hering attacked the Jesuits as the sole cause of the persecution of Protestants in Hungary.

The fact that the appearance of the great comet was immediately followed by a war that lasted for thirty years left a strong impression in people's hearts. In many cases we do not have sufficient primary sources about the common people's reactions to *Wunderzeichen*, but with the 1618 comet, a number of people recorded it in their writings along with their experiences of the dreadful war. For them, the history of the war started from the great comet rather than the defenestration of Prague. Bohemia was too far away, so the comet, seen in local places, was their first direct experience of the war. For example, Joachim Rese, the Lutheran mayor of Jessnitz near Bitterfeld, wrote that a dreadful comet 'whose shape represents a rod of sparks' appeared and remained in the sky for thirty days. He interpreted that this meant the war would continue for thirty years, and 'as unfortunately, the experience proves it sufficiently'.[44] A peasant of Massen near Finsterwalde named Kaschke wrote in his autobiographical writing that he saw the 1618 comet when he was eight years old. Kaschke also stated that it 'shone in the sky for thirty days' and 'the Bohemian [war], and in addition the bloody Swedish-German war followed'.[45] It is clear that this myth of the thirty-day appearance of the comet emerged after the war (Kaschke obviously wrote his account decades later). An anonymous Erfurt chronicle instead focuses on the number 18 surrounding the comet: 'a comet was seen in Erfurt [i]n 1618 on the 18th of November. Thence followed great wars and dying, and it was seen for 18 days in a row'. The reason why the author of the chronicle put prophetic significance to the number is that the author believed that the war had come to an end with the Peace of Prague in May 1635 (he reckoned that the hostilities broke out in 1617).[46] These comments show the strong impression the 1618 comet left in these authors' minds. Samuel Fritz, an Erfurt cook who left a voluminous illustrated chronicle, could vividly recall the shape of the comet that he saw more than forty years ago, when he was eight: 'In the year 1618, this comet stood as a broomstick. Thence followed the Bohemian Unrest, for God swept Germany a lot with the fiery broomstick of His grim Wrath'.[47] Hans Heberle, a pious shoemaker of a village near Ulm, was so shocked by the comet and Dietrich's sermon that he started writing a chronicle of his age. He later wrote:

> Anno 1618 a great comet appeared in the form of a large and horrible rod of punishment, with which God mightily threatened us because of our sinful lives, which we deserved many times over and still deserve today ... what it meant – what also will come of it – that is something we may cry hot tears over, as we, alas, experience now and have experienced from 1620 up to 1630 and which can't described.[48]

Heberle was clearly impressed that Dietrich's prophetic warning became real (though it was a conventional message), and the tone of the Superintendent is echoed in Heberle's writing.

The Decline of *Wunderzeichen* Discourses during and after the War

Since the prophetic warnings of *Wunderzeichen* finally came true with the 1618 comet, we imagine that people would become more attentive to celestial matters. Indeed, the author of an Augsburg broadsheet on the appearance of a yellow rod in the sky in January 1621 opens his text by invoking people's memories of the 1618 comet, which signified 'many wars and conflicts, great disunity in the world'.[49] Obviously the author thought that it would make his text more appealing and persuasive.

However, it is difficult to measure how much the 1618 comet intensified people's interest in *Wunderzeichen* because the number of publications on celestial wonders declined in the following years, and the number of prints never recovered even after the end of the war. What was the cause of this decline? The scope of this study is basically from Luther's death to the outbreak of the Thirty Years' War, but since this question relates to the main point of this study – why *Wunderzeichen* gained such popularity between 1545 and 1618 – here I examine some possible causes of the decline. First, people did not lose interest in *Wunderzeichen* after 1618, and there was no significant change in people's interpretations of celestial phenomena during the war. Since the prediction of war was fulfilled, they had no reason to change their views. People who observed *Wunderzeichen* connected it with the ongoing war. In his chronicle, a Kitzingen pastor recounts how a mysterious fire filled the night sky in January 1630 and states: 'What this dreadful *Feuerzeichen* meant, Germany experienced in the Swedish War for almost 18 years'.[50] The shoemaker Heberle witnessed this same phenomenon, but he also witnessed two armies fighting fiercely in the fiery sky. He adds a simple but impressive comment to his description: 'But what this meant, unfortunately we experience enough and too much, and more than enough'.[51] Considering that Heberle had to flee his village no fewer than thirty times between 1630 and 1648, it is understandable why he wrote thus.[52] These comments were written long after the incident, but people like Heberle may have paid more attention to *Wunderzeichen* after these tough experiences.

The cause of the decline of *Wunderzeichen* literature during the war is not the change in people's views on *Wunderzeichen*. The scarcity of paper and other materials necessary for printing, the general decline of the economy and the severe disruption of trade caused by the war had to be more decisive factors for the decrease in the output of *Wunderzeichen* prints. Moreover, the possible change in people's concerns was also important. The appealing power of the 'repent or disaster' message was strongest when disaster had not yet arrived or existed only in vague reality. Fear and anxiety are strongest when they have not materialized yet. Once the danger of a disaster becomes real and materializes in daily life, people's concerns change. People sought news to judge the present condition of the war and to consider what they should do, rather than making speculations over vague prophetic messages. For example, once the major battlefields moved to Germany, people wanted concrete information about where the carriers of misfortune – troops – were moving, rather than vague prophecies of imminent war, inflation and pestilence. The people already knew these disasters were coming. Like Heberle, they certainly had to have been shocked when wondrous signs appeared in front of them and to speculate on their meanings, but to purchase a broadsheet about strange phenomena that occurred in distant lands was a different matter. Since it was a time of scarcity and economic difficulty, printers had to be more cautious in choosing the titles they should print than in the time before the war.

There is one example that vividly shows the change of the people's mentality about wonder. Flocks of swans suddenly came to the small Thüringen town of Gebesee in late 1634, and it made many local people anxious because it reminded them of the popular expression 'foreign birds bring more foreign guests', which could mean foreign troops. People still remembered that after a great number of wild geese passed through the region two years ago, the imperial army came to their area. The anxious conversations of the locals are recorded in a news tract written by an anonymous author.[53] Unlike the conventional *Wunderzeichen* literature, here the nature of the possible imminent danger is clear and real, and the printer of the tract could expect that it would surely draw the people's attention. What readers of this tract would most likely do is look for information so that they could judge the real possibility of the imminent arrival of the imperial or Swedish troops.

Barnes suggests that a change in attitude towards apocalyptic literature among Lutheran ecclesiastical authorities had to be another significant factor in the decline of *Wunderzeichen* literature. We already saw that during the pre-war period, unorthodox, esoteric and sometimes pansophic ideas crept into Lutheran eschatological discourses. Rosicrucian writings were only a part of them. The Lutheran Church gradually became suspicious of such quests for hidden wisdom and mystical tendencies. Even Arndt was denounced by some prominent theologians because of his emphasis on mystical piety. Thus, as Barnes states,

> [a]dherents of an apocalyptic faith were thus faced with an increasingly clear choice: to maintain a pansophic quest for prophetic wisdom that was now subject to mounting suspicion and discouragement from church leaders, or to abandon the search for gnosis beyond what appeared to many as the lifeless teachings of the church.[54]

In addition, the spread of political prophecy during the war hastened the decline of eschatological literature. The Thirty Years' War caused a flood of Protestant prophecy, and many of these messages insisted that the Habsburgs would be defeated soon and a new golden age would come. Even Jakob Böhme, a shoemaker known for his theosophical writings who had hardly showed deep interest in politics, expressed his hope of a 'new reformation' as the war broke out. Believing that 'the *Aureum Saeculum* will then begin to flourish in the midst of the fire of Babel', he predicted that the nest of 'an eagle' will be taken by lions soon. Of course, this means the downfall of the Habsburgs.[55] However, contrary to these predictions, the course of the war did not bring decisive victory to the Protestant side, and the fighting dragged on endlessly. No messianic emperor appeared. Many people regarded Gustav Adolph of Sweden as the long-awaited 'northern lion' who was supposed to come to save God's people, but his tragic death on the battlefield quickly shattered this dream. Since the eschatological tradition of Lutheranism was the backbone of *Wunderzeichen* discourses, the decline of eschatological literature directly eroded its grounds.

The popularity of *Wunderzeichen* discourses (and also of Lutheran eschatology) did not recover even after peace returned in 1648. This does not mean that people's interest in celestial phenomena had vanished. We can see a number of *Wunderzeichen* broadsheets and other literature published after 1648, and the people's belief that celestial phenomena signified future misfortune remained until the nineteenth century, not only in Germany but in all of Europe. Indeed, the fact that Pierre Bayle, the French pioneer of Enlightenment scepticism, launched his campaign against 'superstition' with his rejection of the portentous interpretation of comets indicates that *Wunderzeichen* belief was rife throughout Europe long after the war.[56] Nonetheless, we can stay that its heyday definitely passed with the outbreak of the Thirty Years' War.

The spread of scientific knowledge, the change of religious mentality represented by the growing individualization of faith among the pietists, and the diversification of literature genres may have contributed to this decline. However, what seems to be important is that the peak period of *Wunderzeichen* discourses (1546–1618) coincided with the development of confessionalization and inter-/inner-confessional conflicts in the Empire. By no means did the process of confessionalization end with the Peace of Westphalia in 1648, but the intense tension among confessional groups and the expectation of apocalyptic battle were eased considerably. Calvinism was recognized as a legal religion, and the relentless advance of a militant Counter-Reformation was contained. Now each territory could concentrate on their own state-building. Theologically, Lutheran orthodoxy was firmly established, and even ardent pietists did not challenge the Book of Concord.

Conclusion

The decade prior to the war was filled with growing political and confessional tensions between Lutherans, Calvinists and Catholics. In this political crisis, the possibility of 'change of regime', 'disunity among princes' and 'death of princes', the familiar warnings of *Wunderzeichen* discourses, became deadly real. It is understandable that many *Wunderzeichen* discourses published during this period were filled with anxiety. However, despite the dramatic change of political conditions, there was no substantial change in the contents of *Wunderzeichen* discourses themselves. The proto-pietism of Arndt and the Rosicrucian ideas did not affect the basic tenets of *Wunderzeichen* messages because these trends maintained the framework of the traditional eschatological world view of Lutheranism.

The lack of the Gospel aspect in *Wunderzeichen* messages became clear as the significance of trusting in God's caring love was emphasized in pastoral care in the early seventeenth century. Compared with stories of angelic encounters, reports of celestial wonders were less assuring about God's intervention in human affairs. Despite these limitations, *Wunderzeichen* remained as a popular genre of religious literature throughout this stormy period.

The publication of *Wunderzeichen* literature started declining during the Thirty Years' War, and the trend continued even after the end of the war. This was not caused by the change of people's views on celestial phenomena. Having seen the fulfilment of the 1618 comet's prophetic message, people kept a keen interest in strange signs. Several factors for this decline can be considered, but there are two important points: first, the Lutheran eschatological world view was no longer a dominant element of faith, and since eschatology was the backbone of *Wunderzeichen* discourses, these discourses lost their power and attraction. Moreover, the heyday of *Wunderzeichen* discourses coincides with the age of confessionalization, and as the struggle for confessional hegemony came to an end with the Peace of Westphalia, the clergy's passion for *Wunderzeichen* also declined.

CONCLUSION

We have explored the world of sixteenth-century *Wunderzeichen* discourses, focusing on the interaction between celestial wonders, confessional conflicts, apocalypticism and the perception of human sin. Let us summarize what we have discovered from these sources, particularly the Lutheran clergy's publications on *Wunderzeichen*.

People in the sixteenth century showed great interest in comets, halo phenomena, *Fewerzeichen* and various celestial visions, which they regarded as signs of God's wrath and impending punishment. Many Lutheran theologians and pastors shared this interest in celestial phenomena, actively engaging in the production of *Wunderzeichen* literature. The list of theologians who discussed *Wunderzeichen* in their writings includes prominent ecclesiastical leaders from the latter half of the sixteenth century, such as all five formulators of the Formula of Concord. There were several factors that encouraged their interest in *Wunderzeichen*. The foremost factor was their eschatological conviction that the Day of the Last Judgement was at hand and the final countdown to the End Times had started with Luther's Reformation. The Gospel and the Book of Revelation warned that restoration of the Gospel, the revelation of Antichrist and the persecutions of believers would occur prior to the Last Judgement, and Lutheran theologians generally agreed that all of these prophecies had come true during their days. Various celestial phenomena strengthened their eschatological conviction because Christ explicitly stated that various signs would appear among the heavenly bodies.

In addition, the clergy found that *Wunderzeichen* discourses were useful tools for their ministry. Since *Wunderzeichen* were commonly regarded as signs of God's wrath, they could be used as *Gesetzpredigt*, preaching of the Law. In the Lutheran pedagogy of conversion, one must first be terrified by the demand of the Law, which shows God's wrath on sinners. Only after one recognizes his or her powerlessness to achieve salvation by one's own power, can one find the true meaning of the Gospel. Thus the clergy incorporated shocking reports of *Wunderzeichen* in their sermons to stress God's wrath and the necessity of repentance. Moreover, in order to make hearers recognize their sinfulness and to urge them to consider their spiritual well-being more seriously, they filled their messages with harsh accusations of moral failure in people. This kind of jeremiad became a familiar part of *Wunderzeichen* discourses.

Thus, as Paul Eber has stated, *Wunderzeichen* were *Beiprediger*, the co-workers of the Lutheran preachers. Many clergy stated that since people were not willing to listen to God's Word conveyed through their sermons, God placed wondrous signs as 'another preacher' and let them preach repentance from the sky. This idea of *Beiprediger* reflects the Lutheran clergy's frustration with their parishioners. During the Reformation, the clergy lost the spiritual authority that the Catholic priests had – the sacred power of transubstantiation. Now the only foundation of the clergy's authority and identity was God's Word. However, for many clergymen, the people's zeal for God's Word (and respect for God's servants) was far from adequate. Many pastors had to struggle and negotiate with parishioners for their authority as God's agents in the community. For them, the appearances of *Wunderzeichen* – which shocked people and made their hearts more receptive to preachers' warnings, at least temporarily – offered a good chance to reassert the authority of God's Word and to warn against the neglect of God's Word (and God's servants).

Some theologians tried to exploit the popularity of *Wunderzeichen* discourses for more specific confessional agendas. As the Interim Crisis broke out, Matthias Flacius Illyricus published an account of the celestial visions witnessed by Braunschweig merchants as part of his anti-Interim campaign. The mysterious visions of an eagle wounding himself, a weeping maid and the image of the 'Martyr Elector' Johann Friedrich made the news also a convenient propaganda tool against imperial religious policy and concessions by the Wittenberg theologians.

The champion of *Wunderzeichen* storytellers among the Lutheran clergy was Christoph Irenaeus. From the early years of his career as a pastor and a relentless polemicist, he expressed his profound interest in *Wunderzeichen* and used stories he had collected to attack his theological opponents. His *Wasser-Spiegel* (1566), a work that focused on floods and storms, established Irenaeus's basic strategy: to draw readers' attention through various stories of wonder and lead them to his theological agenda. For Irenaeus, strange signs and natural disasters were manifestations of God's wrath against false teachers such as Synergists and Calvinists. During the 1570s and the 1580s he turned his aim against Jacob Andreae and supporters of the Formula of Concord because he could not accept Andreae's understanding of original sin and human nature, as incorporated in the Formula. Irenaeus focused on comets and other celestial *Wunderzeichen* in the *Prognosticon Aus Gottes Wort* and used these stories of wonder as an introduction to his refutation of the Concord project and of Andreae's anti-Flacian position. In his last *Wunderzeichen* book, *De Monstris*, Irenaeus connected the image of monsters with his Flacian understanding of the 'monstrous' nature of humankind after the Fall. He also transformed the stories of births of double-tongued babies into the introduction of his attack against the double-tongued attitude of his theological opponents. Throughout Irenaeus's *Wunderzeichen* books, four major issues repeatedly appear: *Wunderzeichen*, false preachers, persecutions of believers and the imminent coming of the Last Judgement. In the heart of the *Exul Christi* Irenaeus, who repeatedly suffered expulsions due to his theological positions, these four points were inseparably intertwined, and they strengthened his conviction that Christ's return was at hand.

What we can learn from these propagandistic uses of *Wunderzeichen* is that for these authors (especially Irenaeus), theological controversies had cosmic significance that moved God to show special signs in the skies. Strange celestial phenomena endorsed the theologians' own positions and revealed the error of their opponents' views. However, not all theologians expressed interest in *Wunderzeichen* with the same degree of enthusiasm. Andreae mentioned *Wunderzeichen* in his writings several times, but it seems that he was hardly enthusiastic about this popular topic. Nonetheless, Andreae knew that *Wunderzeichen* had a strong appeal among the common people, and out of his pastoral concern, he did not hesitate to use stories of wonder when it seemed useful to edify the laity. Thus Andreae used the astrological imagery of the 'five planets' to admonish people not to indulge in vices, although he himself had no interest in astrology. In his Weimar sermon, he quoted the recent news of the great comet, but it served only to make his message more attractive to his audience/readers. His arguments in this sermon suggest that the reason why Andreae did not discuss *Wunderzeichen* often in his writings was that he was worried about the possibility that careless references to celestial phenomena would encourage people's interest in astrology. This cautious attitude reflects Andreae's genuine concern about lay morality and shows that the Late Reformation Lutheran sermons were not dry, dogmatic messages unrelated with the practical spiritual needs of the laity.

Although the political situation of the Lutheran Church rapidly changed in the last two decades prior to the outbreak of the Thirty Years' War, the basic tenets of *Wunderzeichen* did not change. During this period, two new trends appeared in Lutheran spirituality: proto-pietism and unorthodox esoteric faith. However, since these new trends still retained the traditional eschatological world view, they did not add new elements to *Wunderzeichen* discourses. At the same time, the increase of sermons that placed an emphasis on spiritual consolation and encouragement in the early seventeenth century reveals the limitation of the *Wunderzeichen* message's appeal. Discourses on celestial wonders also lacked the personal touch of the reports of angelic encounters, and it is clear that *Wunderzeichen* stories could not effectively teach people about God's caring love.

The repeated predictions of imminent disaster were finally fulfilled when the Thirty Years' War broke out after the appearance of the great comet in

November 1618. However, this was ironically the beginning of the decline of *Wunderzeichen* discourses. People's interpretation of celestial phenomena did not change, but the scarcity of raw materials and the disruption of trades caused by the war led to a decrease in the publication of *Wunderzeichen* literature. Moreover, facing the real danger of war in their lives, people sought real news of the war rather than vague warnings of celestial phenomena. This shows that prophecies of disaster have the strongest impact when the disaster has not yet arrived. Once people's fears are materialized, they no longer need prophecy; they have to prevent the situation from becoming worse by their immediate practical actions. In addition, during this period, church authorities grew suspicious of chiliastic prophecies about the apocalyptic time schedule and the (sometimes political) expectation of a golden age. Since eschatology was the backbone of *Wunderzeichen* discourses, the authorities' negative attitude towards eschatological literature affected the production of *Wunderzeichen* works profoundly.

The above examination shows that the heyday of *Wunderzeichen* discourses was the period between Luther's death (1546) and the outbreak of the Thirty Year's War (1618). This timeframe coincides with the period of confessionalization, the solidification and development of a Lutheran identity in theological, political and social terms. It can roughly be divided into two phases. First, during the period between Luther's death and the formulation of the Formula of Concord (1577), Lutheran theologians struggled to define genuine Lutheran theology and who the true heirs of Luther were. Theologically, this process of solidification of Lutheran orthodoxy was completed with the publication of the Book of Concord in 1580, although several theologians like Irenaeus never accepted such agreement. In the second phase leading to the fatal year of 1618, opposition to the Concord gradually died out, which is why the death of Irenaeus in 1595 is used as a chronological boundary. However, it did not stabilize the political and social conditions surrounding the Lutheran Church. People's moral failures made church leaders suspect that the restoration of the Gospel had failed to change people's hearts, and they repeated the same lamentation that 'although the Gospel has been preached for long years, we find no improvement in the people'. In addition, the years following the Concord were characterized by growing tensions between the three confessional camps: Lutherans, Calvinists and Roman Catholics. Throughout this period, Lutherans had to confront a rapid advance of Calvinist influences and Tridentine Counter-Reformation programmes, and this mounting confessional tension finally exploded with the outbreak of the war.

Lutheran theologians projected their concerns about the condition of their Church and people's moral behaviours onto celestial phenomena. They believed that God would surely express his wrath against the people's moral failures and the hideous errors of false prophets. Christ's prophecy of the End Times – celestial wonders, a decline of morality and the spread of erroneous teachings – gave theological foundation to their convictions. It is no wonder that the clergy's interest in *Wunderzeichen* was most intense during this age of confessionalization. And as Barnes's study shows, this period was also the heyday of Lutheran apocalypticism. Irenaeus's connections between *Wunderzeichen*, false teachers, religious persecutions and the Last Judgement were not just due to his eccentric personal obsessions. These four factors were in fact major concerns for the clergy living in the period of confessionalization, and tightly connected to their eschatological conviction. The Peace of 1648 did not bring an end to the process of confessionalization, but it defined the geographical boundaries of confessional influence and released the political and religious tension between confessional camps. This may be one important factor in the quiet decline of *Wunderzeichen* literature and eschatological discourses in the latter half of the seventeenth century.

Wunderzeichen discourses reveal interesting aspects of early modern religious culture that we cannot learn from other sources. They reveal the dynamic interconnections between nature, print culture and the confessionalization process. Prints defined people's way of observing the world of nature and directed what kind of impression they would receive from their observation. Through the clergy's apocalyptic *Wunderzeichen* publications, celestial phenomena were confessionalized and became active tools to promote specific agendas and strengthen people's religious identity. In this sense, we can say that sixteenth-century Lutheran apocalypticism was essentially a media event. Ironically, this printed apocalypticism was later overwhelmed by a real apocalyptic catastrophe, the Thirty Years' War. However, *Wunderzeichen* discourses could at least prove the validity of their countlessly repeated messages before their final decline: the bright, 'broomstick'-shaped comet in 1618, which achieved a mythical status for those that survived the destructive fury of war.

NOTES

Introduction: Celestial Wonders, Confessional Conflicts and Apocalypticism

1. M. Schwegler, *'Erschröckliches Wunderzeichen' oder 'natürliches Phänomenon'?: Frühneuzeitliche Wunderzeichenberichte aus der Sicht der Wissenschaft* (Munich: Percy Berktold, 2002).
2. G. Hellmann, *Die Meteorologie in den deutschen Flugschriften und Flugblättern des XVI Jahrhunderts* (Berlin: Preussischen Akademie der Wissenschaften, 1921). Hellmann also lists more than 100 prints published between 1500 and 1549.
3. B. Weber, *Wunderzeichen und Winkeldrucker 1543–1586* (Zurich: Urs Graf, 1972).
4. W. L. Strauss (ed.), *The German Single-Leaf Woodcut, 1550–1600*, 3 vols (New York: Abaris Books, 1977).
5. D. Alexander (ed.), *The German Single-Leaf Woodcut, 1600–1700*, 2 vols (New York: Abaris Books, 1977).
6. W. Harms (gen. ed.), *Deutsche Illustrierte Flugblätter des 16. und 17. Jahrhunderts* (Tübingen and Munich, 1980–). The following volumes contain *Wunderzeichen* broadsheets: W. Harms, M. Schilling, B. Bauer and C. Kemp (eds), *Die Sammlung der Herzog August Bibliothek in Wolfenbüttel, Teil 1. Ethica and Physica*, Volume 1 of the series (Tübingen: Max Niemeyer, 1985); W. Harms and C. Kemp (eds), *Die Sammlungen der Hessischen Landes- und Hochschulbibliothek in Darmstadt*, Volume 4 of the series (Tübingen: Max Niemeyer, 1987); W. Harms and M. Schilling (eds), *Die Sammlung der Zentralbibliothek Zürich Teil 1: Die Wickiana I (1500–1569)*, Volume 6 of the series (Tübingen: Max Niemeyer, 2005); and W. Harms and M. Schilling (eds), *Die Sammlung der Zentralbibliothek Zürich Teil 2: Die Wickiana II (1570–1588)*, Volume 7 of the series (Tübingen: Max Niemeyer, 1997).
7. C. Hoffmann-Randall (ed.), *Monster, Wunder und Kometen: Sensationsberichte auf Flugblättern des 16. bis 18. Jahrhunderts* (Erlangen: Universitäts Bibliothek, 1999).
8. A. Janeck (ed.), *Zeichen am Himmel: Flugblätter des 16. Jahrhunderts* (Nuremberg: Germanisches National Museum, 1982).
9. F. Mauelshagen, 'Illustrierte Kometenflugblätter in wahrnehmungsgeschichtlicher Perspektive', in W. Harms and M. Schilling (eds), *Das Illustrierte Flugblatt in der Kultur der Frühen Neuzeit* (Frankfurt am Main: Peter Lang, 1998), pp. 101–36; A. Messerli, 'Angst und Wunderzeichen in Einblattdrucken', *Librarium*, 31 (1988), pp. 182–97; M. Schilling, 'Flugblatt und Krise in der Frühen Neuzeit', in W. Harms and A. Messerli (eds), *Wahrnehmungsgeschichte und Wissensdiskurs im illustrierten Flugblatt der Frühen Neuzeit (1450–1700)* (Basel: Schwabe, 1999), pp. 33–56; S. Homeyer, 'Die

kompensatorische Funktion für den Rezipenten bei einigen illustrierten Flugblättern des 16. Jahrhunderts mit Endzeitvorstellungen', in W. Harms and M. Schilling (eds), *Das Illustrierte Flugblatt in der Kultur der Frühen Neuzeit* (Frankfurt am Main: Peter Lang, 1998), pp. 137–49; and S. Homeyer, '"...das ende mus verhanden sein!...": Studien zur eschatologischen Bildlichkeit auf illustrierten Flugblättern der Frühen Neuzeit' (PhD dissertation, Otto-von-Guericke University, 2002). Michael Bischoff's book, for popular reading, also contains many woodcuts from broadsheets: *Himmelszeichen: Eine bildreiche Kunde von Aberglauben und Ängsten* (Nördlingen: Franz Greno, 1986).

10. A. Walsham, 'Sermons in the Sky: Apparitions in Early Modern Europe', *History Today*, 51:4 (2001), pp. 56–63, on pp. 57–8.
11. R. Schenda, 'Die deutschen Prodigiensammlungen des 16. und 17. Jahrhunderts', *Archiv für Geschichte des Buschwesens*, 4 (1963), pp. 637–710; R. Schenda, 'Wunder-Zeichen: Die alten Prodigien in neuen Gewändern: Eine Studie zur Geschichte eines Denkmusters', *Fabula*, 38 (1997), pp. 14–32; H. Schilling, 'Job Fincel und die Zeichen der Endzeit', in W. Brückner (ed.), *Volkserzählung und Reformation: Ein Handbuch zur Tradierung und Funktion von Erzählstoffen und Erzählliteratur im Protestantismus* (Berlin: Erich Schmidt, 1974), pp. 326–93; and B. Deneke, 'Kaspar Goltworm: Ein lutherischer Kompilator zwischen Überlieferung und Glaube', in W. Brückner (ed.), *Volkserzählung und Reformation: Ein Handbuch zur Tradierung und Funktion von Erzählstoffen und Erzählliteratur im Protestantismus* (Berlin: Erich Schmidt, 1974), pp. 125–77.
12. Volker Leppin's *Antichrist und Jüngster Tag: das Profil apokalyptischer Flugschriftenpublizistik im deutschen Luthertum* (Gütersloh: Gütersloher Verlagshaus, 1999) discusses the apocalyptic elements of *Wunderzeichen* news on pp. 87–96.
13. P. M. Soergel, *Miracles and the Protestant Imagination: The Evangelical Wonder Book in Reformation Germany* (Oxford: Oxford University Press, 2012).
14. C. S. Dixon, 'Popular Astrology and Lutheran Propaganda in Reformation Germany', *History*, 84 (1999), pp. 403–18, on p. 414.
15. C. Methuen, *Kepler's Tübingen: Stimulus to a Theological Mathematics* (Aldershot: Ashgate, 1998), p. 7.
16. Ibid., pp. 73–7.
17. Ibid., pp. 132-6.
18. Ibid., pp. 129–32.
19. Ibid., p. 208.
20. R. B. Barnes, *Prophecy and Gnosis: Apocalypticism in the Wake of the Lutheran Reformation* (Stanford, CA: Stanford University Press, 1988), p. 2.
21. Ibid., p. 87.
22. Ibid., p. 170.
23. Ibid., p. 65.
24. R. K. Rittgers, *The Reformation of Suffering: Pastoral Theology and Lay Piety in Late Medieval and Early Modern Germany* (Oxford: Oxford University Press, 2012), p. 191.
25. J. Spinks, *Monstrous Births and Visual Culture in Sixteenth-Century Germany* (London: Pickering & Chatto, 2009), p. 59.
26. Ibid., p. 146.
27. J. Crawford, *Marvelous Protestantism: Monstrous Births in Post-Reformation England* (Baltimore, MD: Johns Hopkins University, 2005), p. 109.
28. Ibid., p. 126, p. 139.
29. Ibid., p.76.
30. Soergel, *Miracles*. p. 30.

31. Ibid.
32. Ibid., p. 128.
33. A. Walsham, *Providence in Early Modern England* (Oxford: Oxford University Press, 1999); J. Friedman, *The Battle of the Frogs and Fairford's Flies: Miracles and the Pulp Press during the English Revolution* (New York: St Martin's Press, 1993); J. Spinks, 'Print and Polemic in Sixteenth-Century France: The *Histories prodigieuses*, Confessional Identity, and the Wars of Religion', *Renaissance Studies* (2011), pp. 1–25.

1 Exploring the World of *Wunderzeichen*

1. C. Spangenberg, *Mansfeldische Chronica: Der Erste Teil* (Eisleben, 1572).
2. Weber, *Wunderzeichen und Winkeldrucker*, p. 30.
3. *Drey Sonnen: wie dieselben mit mancherley Regenbögen zu Witeberg vnd weit herumb an der Elb sind lenger denn anderhalb stund gesehen worden* (Wittenberg, 1551); in Harms and Schilling (eds), *Wickiana I*, p. 99 (plate 48).
4. *Warhafftiger Bericht von dem Cometen welcher October dises LXXX. Jars erstlich erschinen vnd noch am Himmel zusehen ist* (Augsburg, 1580).
5. *Schröckliche Newe Zeitung von dem Wunderzeichen welches den kutz verschinenen fünfften dess Christmonats zu Alttorff inn dem Land Württenberg ist gesehen worden.* (Strassburg, 1578); in Harms and Schilling (eds), *Wickiana I*, p. 79 (plate 38).
6. *Ein Erschrecklich vnd Wunderbarlich zeychen so am Sambstag für Judica den zehenden tag Martij zwishcen siben vnd acht vhrn in der Stadt Schalon in Franckrey von vielen leuten gesehen worden* (Nuremberg, 1554). Strauss (ed.), *The German Single-Leaf Woodcut*, vol 1, p. 416; Harms and Schilling (eds), *Wickiana I*, p. 125 (plate 61).
7. *Wunderzeichen so zu grossen Sara ein meil vonn Gera einem Stettlin im Voytland gelegen den neundten Martij am morgen vmb siben vhr diss 1563. jars am hellen Himmel gesehen worden* (Frankfurt am Main, 1563); in Harms and Schilling (eds), *Wickiana I*, p. 251 (plate 125).
8. Schwegler, *Erschröckliches Wunderzeichen*, p. 30.
9. W. Pletzlein, *Warhafftige Bericht Vnd Kurtze Beschreibung des Grossen vnnnd Erschrecklichen Wunderzeichens So sich am heiligen Ostertag den 19. Aprils dieses jtzlauffenden 84. Jars an der Sonnen begeben* (Hof, 1584).
10. Barnes, *Prophecy and Gnosis*, p. 87.
11. H. Fincel, *Wunderzeichen* (Jena, 1556).
12. C. Lyscothenes, *Wunderwerck Oder Gottes vnergründliches vorbilden* (Basel, 1557).
13. C. Goltwurm, *Wunderwerck vnd Wunderzeichen Buch* (Frankfurt am Main, 1557).
14. S. Suevus, *Cometen: Was sie für grosse Wunder vnd schreckliche ding zu bedeuten vnd anzukündigen pflegen* (Görlitz, 1578); *Spiegel des Menschlichen Lebens* (Leipzig, 1588). This is reprinted in S. Suevus, *Erbauungsschriften: Spiegel des Menschlichen Lebens*, ed. M. A. van den Broek (Amsterdam: Rodopi, 1984).
15. G. Caesius, *Chronick Oder ordenliche verzeichnuss vnnd beschreibung aller Cometen* (Nuremberg, 1579).
16. J. Hebenstreit, *Prognosticon Historicvm Auffs 1568. Jhars* (Erfurt, 1567).
17. J. Wittich, *Gründtliche vnd warhafftige Bericht von dem Erschrocklichen vnd Wunderbarlichen Zeichen* (Erfurt, 1561), fol. A1v.
18. A. Ursinus, *Kurtze Beschreibunge der geschehen vnd gesehenen vnnatürlichen Wunderzeichen am Himmel, im 1568. 69. vnd 70. Jhare* (Erfurt, 1570), fol. B1r.
19. 'Exorcism', in *Ten Colloquies of Erasmus*, ed. C. R. Thompson (New York: Liberal Arts Press, 1957), p. 39.

20. *Warhafftige beschreibung: was auff einen jeden sollichen Cometen geschehen sey die gesehen sind von anfang der Welt her biss auff disen ietz geschenen Cometen in dem 56. Jar* (Strasbourg, 1556); in Harms and Schilling (eds), *Wickiana I*, p. 153 (plate 75).
21. J. Coler, *Eigentlicher bericht vo[n] den seltzamen vnd zu vnserer Zeit vnerhörten Wunderwercken vnd Geschichten so sich newlicher zeit in der Marck Brandenburg zugetragen vnd verlauffen haben vnd noch teglich geschehen* (Erfurt, 1595), fols B4r–B4v.
22. Anon., *Warhafftige newe Zeitung vnd erschreckliche Wunderzeichen so sich newlichen zu Rom vnnd zu Paris in dem nechst vergangenen 83. Jar geschehen* (Augsburg, 1584).
23. Schwegler, *Erschröckliches Wunderzeichen*, p. 62.
24. Caesius, *Chronick*.
25. B. Gernhard, *Vom Jüngste Tage: Vier nützliche Predigten inn Zwölff Heuptartickel* (Erfurt, 1556), fol. B3v.
26. On the controversy over the nature of the 1572 nova in Germany, see C. Methuen, '"This Comet or New Star": Theology and the Interpretation of the Nova of 1572', *Perspectives on Science*, 5 (1997), pp. 499–514.
27. Aristotle, quoted in J. L. Jervis, *Cometary Theory in Fifteenth-Century Europe* (Dordrecht: D. Reidel, 1985), p. 12.
28. Ibid., p. 13.
29. Suevus, *Erbauungsschriften*, fol. b1r.
30. N. Orphanus, *Beschreibung des Erschrecklichen Brennenden Flammenden vnd Stralschiessenden Fewers vnd Zornzeichen Gottes vber Europa* (Eisleben, 1575), fols D2v–D3r.
31. Ibid., fol. D3r.
32. Caesius, *Chronick*, fol. S6r.
33. J. Pilich, *Eine Predigt Am newen Jaarstage mit kürtzer vermanung zur Busse vnd bekerung* (Magdeburg, 1581), fol. B3r.
34. For example, *Newe Zeitung: Von dem wunderbaren neuwen vnnd vor nicht mehr gesehenen Sternen* (Lauingen, 1573); in Harms and Schilling (eds), *Wickiana II*, p. 83 (plate 43).
35. A. Ursinus, *Prognostication Auff das Jhar nach der Geburt Jhesu Christi vnsers Heylandes. M.D.LXXIIII: Beyneben einer kurtzer Beschreibung des erschienen Cometens im 1572. vnd 1573 Jhare* (Erfurt, 1574), fol. D1r.
36. The mechanism of these halo phenomena is explained in J. Naylor, *Out of the Blue: A 24-Hour Skywatcher's Guide* (Cambridge: Cambridge University Press, 2002), pp. 136–56; and M. Maunder, *Lights in the Sky: Identifying and Understanding Astronomical and Meteorological Phenomena* (London: Springer, 2007), pp. 89–92, pp. 161–74.
37. Ursinus, *Kurtze Beschreibunge*, fols B2v–B3r.
38. Ibid., fols B3r–B3v.
39. Hebenstreit, *Prognosticon Historicvm*, fol. B3r.
40. J. Hebenstreit, *Wunderzeichen* (Erfurt, 1564).
41. *Warhaffte Contrafactur derer jüngst erschienen grossen Wunderzeychen dreyer Sonnen vier Regenbogen vnd darinn zweyer Liechter klarheyten auch grossen weiss Creutzes am Himel an villen orten gesehen worden* (Nuremberg, 1580); in Strauss (ed.), *The German Single-Leaf Woodcut*, vol. 2, p. 656.
42. W. Peristerus, *Eine Trostreiche Predigt Vber das Euangelium Vom Cananeishcen Weiblein ... Sampt einer kurtzen Erinnerung vnd Erklerung deß schrecklichen Wunderzeichen so den Sonabend zuuor nach Mittage erschienen vnnd gesehn ist worden* (Berlin, 1581), fols E2v–E3r.
43. Two broadsheets (1571), which report that three suns and three strange rainbows appeared in the sky over Cologne in January, regard the phenomenon as a sign of immi-

nent Judgement. See *Warhafftige doch erschröckliche Gesicht so gesegen ist worden an dem Himel zu Coln am Rhein den 26. Januarij des 1571 Jars* (Augsburg, 1571), in Strauss (ed.), *The German Single-Leaf Woodcut*, vol. 2, p. 753; and *Jm disem Anno MD.LXXI. Jar am XXV tag Ianuarij. sindt zu Cöln am Rein drei Sonnen gesehe[n] worden* (Frankfurt am Main, 1571), in Strauss (ed.), *The German Single-Leaf Woodcut*, vol. 1, p. 136.

44. Peristerus, *Trostreiche Predigt*, fols E3v–E4r.
45. Scientists point out that 'the great auroras that appear at low latitudes are also more likely to be red, which gives them an even more threatening appearance'; C. Savage, *Aurora: The Mysterious Northern Lights* (San Francisco, CA: Sierra Club Books, 1994), p. 37.
46. Spangenberg, *Mansfeldische Chronica*, fol. 480r.
47. G. Licht, *Von dem schröcklichen vnd betrübtem Nawen* [*sic*] *Jhar so an den Wolcken des Himels den 38.* [*sic*] *Decembris dieses 61. Jhars gesehen* (Frankfurt an der Oder, 1561), fol. A4v.
48. Orphanus, *Beschreibung*, fols A3v–A4r.
49. J. Baumgarten, *Beschreibung Der Schrecklichen Brunst Fewersklufft Chasmatis vnd Gesichtes so den 27 Augusti des jetzt lauffenden 1581. Jars* (Magdeburg, 1581), fol. B2v.
50. Hebenstreit, *Prognosticon Historicvm*, fols C2v–C3r.
51. H. Fincel, *Der dritte theil Wunderzeichen* (Frankfurt am Main, 1567), fols Ff8v–Gg1r.
52. Baumgarten, *Beschreibung*, fols C4v–D1r.
53. Ibid., fols D2v–D3r.
54. J. Cuno, *Erinnerung Von dem schrecklichen Fewerschuss Welcher am andern Tages des Herbsmonats früe Morgens zwischen 2. Vnd 3. Vhr gesehen worden* (Eisleben, 1579), fol. A7v.
55. Schwegler, *Erschröckliches Wunderzeichen*, pp. 82–3.
56. J. Schütz, *Von den Wunderzeichen am Himel vor dem grossen tage des Herr aus dem Sprüch des Propheten Joel am andern Capittel* (Wittenberg, 1582), fol. A2v, fol. C1r.
57. Fincel, *Wunderzeichen*, fol. a2v. In addition, the strange apparition that Christoph Irenaeus witnessed in the night sky, in February 1564, neatly fits Schwegler's view in *Erschröckliches Wunderzeichen*. While he was observing *Fewerzeichen*, Irenaeus saw two armies of soldiers come out from the clouds, fight each other, and then vanish into a pool of sulphur-like smoke. Since the apparition was seen in the midst of *Fewerzeichen*, it might have been an illusion created by the aurora.
58. *Jm M. D. LIIII. Jar den XI. Tag Brachmonats ist diss gesicht oder zeychen zum Blech fünff meyl von Nürmberg gelegen von vilen menschen gesehen worden* (Strasbourg, 1554). Harms and Schilling (eds), *Wickiana I*, p. 127 (plate 62). Fincel, *Wunderzeichen*, fols b3r–b3v.
59. Walsham, 'Sermons in the Sky', pp. 57–8.
60. Fincel, *Wunderzeichen*, fols c3r–c3v.
61. Ibid., fols Z7r–Z7v, d5v.
62. C. Schlüsselburg, *Postilla* (Frankfurt am Main, 1604), p. 46.
63. *Ein erschröcklich gesicht so zu Embßkirchen auff Erichtag den vierdten tag Marcij bey nacht an dem Hymel gesehen worden* (Augsburg, 1561); in Harms and Schillig (eds), *Wickiana I*, p. 227 (plate 113).
64. G. Hirschmann (ed.), *Die Kirchenvisitation in Landgebiet der Reichsstadt Nuernberg 1560 und 1561: Quellenadition* (Neustadt an der Aisch: Degener, 1994), p. 202.
65. *Wunderbare schröckliche Newe Zeitunge Von dreien mercklichen Himelszeichen* (Strasbourg, 1574); in Strauss (ed.), *The German Single-Leaf Woodcut*, vol. 1, p. 186.
66. Fincel, *Der dritte theil Wunderzeichen*, fol. Ff7v.

67. J. Schütz, *Newe Zeytung: Was sich allhie im Landt zu Sachssen hat zugetragen in der Stadt Bitterfeldt an an [sic] der Dam wie ein menschliche Handt mit einem Blut rhoten Schwerdy am Hellen Himel gestanden ist vnd Blut vom Himel gefallen ist. Anno 1568* (n.p., 1568).
68. Wittich, *Gründtliche*, fol. A3v.
69. *Erschrecklichen Wunderbarliche vmmd warhafftige mirakel vnd zeychen inn machgeschrieben Stetten vnd Flecken grausam ersehen in vergangen vnnd gegenwertigen Jar vnsers Herrn M.D. LIII. Vnd M.D.LV.* (Erfurt, 1555), fol. A2v.
70. Fincel, *Wunderzeichen*, fol. Q8r.
71. Anon., *Ein Gesicht am Himel gesehen: Wie das Götliche Gericht Am tag Visitations Anno 1556* (n.p., 1556).
72. C. von Heisterbach, *The Dialogue on Miracles* (London: Routledge, 1929), vol. 2, pp. 306–7.
73. Hermann Vischer the Younger's drawing of this demonic horse-riding procession is printed in C. Zika, *The Appearance of Witchcraft: Print and Visual Culture in Sixteenth-Century Europe* (London and New York: Routledge, 2007), p. 110.
74. C. Pfister and R. Bràzdil, 'Climatic Variability in Sixteenth-Century Europe and its Social Dimension: A Synthesis', in C. Pfister, R. Brazdil and R. Glaser (eds), *Climatic Variability in Sixteenth-Century Europe and its Social Dimension* (Dordrecht: Kluwer, 1999), pp. 5–53.
75. W. Behringer, 'Die Krise von 1570: Ein Beitrag zur Krisengeschichte der Neuzeit', in M. Jakubowski-Tiessen and H. Lehmann (eds), *Um Himmels Willen: Religion in Katastrophenzeiten* (Göttingen: Vandenhoeck and Ruprecht, 2003), pp. 51–156, on p. 77.
76. J. Pfeffinger, *Christliche Gewisse Deutung der Zeichen die für vnd in diesem 1562. Jar gesehen* (Leipzig, 1562), fols B1v–B2r.
77. Spangenberg, *Mansfeld Chronica*, fol. 485r.
78. P. Victorius, *Xasmatodogia Oder Erinnerung von der schrecklicher Chasmate oder Fewerzeichen* (Magdeburg, 1583), fol. D1r.
79. Behringer, 'Die Krise von 1570', p. 82, p. 90.
80. Ibid., p. 100.
81. Spangenberg, *Mansfeld Chronica*, fols 496vr–501v.
82. Behringer, 'Die Krise von 1570', p. 85.
83. Ibid., pp. 92–3.
84. Pfister and Brazdil, 'Climatic Variability', p. 19, p. 24.
85. R. Brazdil et al., 'Flood Events of Selected European Rivers in the Sixteenth Century', in C. Pfister, R. Brazdil and R. Glaser (eds), *Climatic Variability in Sixteenth-Century Europe and its Social Dimension* (Dordrecht: Kluwer, 1999), pp. 252–61.
86. W. Behringer, *Witchcraft Persecutions in Bavaria* (Cambridge: Cambridge University Press, 1997), p. 94.
87. W. Behringer, 'Climatic Change and Witch-Hunting: The Impact of the Little Ice Age on Mentalities', *Climatic Change*, 43 (1999), pp. 335–51.
88. Caesius, *Chronick*, fol. S6r.
89. A. Plakans, *Historical Dictionary of Latvia* (Lanham, MD: Scarecrow Press, 1997), pp. 108–9.
90. For example, Anon., *Sehr grewliche erschrockliche vor vnerhörte warhafftige Newe Zeytung: was für grausame Tyranney der Moscowiter an den Gefangenen hinweggefürten Christen auss Lyffland* (Nuremberg, 1561).
91. Hebenstreit, *Prognosticon Historicvm*, fol. C3r.

92. J. Renner, *Johannes Renner's Livonian History 1556–1561*, trans. J. S. Smith and W. Urban (Lewiston, NY: Edwin Mellen Press, 1997), p. 19.
93. C. Fagius, *Wunder Stern vnd Zornzeichen: So an Sonn vnd Monde des 1568. Jars den. 14. 21. vnd 22. Tag Decembris zu Erffordt vnd angrentzenden örtern gesehen worden* (Erfurt, 1568), fols B3r–B3v.
94. This is from Nicholas Selnecker's recommendation of Andreas Celichius's book, printed in A. Celichius, *Christliche Notwendige nutzliche vnd Theologische erinnerung von dem newen Cometen* (Leipzig, 1578), fol. A2v.
95. G. J. Miller, 'Holy War and Holy Terror: Views of Islam in German Pamphlet Literature, 1520–1545' (PhD dissertation, Boston University, 1994), pp. 67–8.
96. The history of the conflicts between the Ottomans and the Hapsburgs in the sixteenth century is summarized in M. S. Anderson, *The Origins of the Modern European State System, 1494–1618* (London: Longman, 1998), pp. 231–51.
97. Miller, 'Holy War and Holy Terror', p. 1.
98. Fincel, *Der dritte theil Wundezeichen*, fol. Ff2r.
99. *Drey Newe Zeytung: ... Die ander Zeittung lautet von dreyen Sonnen an dem Himmel sampt etlichen Leichten vnnd anderm Kriegsvolck so gesehen worden im Land zu Poelen* (Erfurt, 1581), in Strauss (ed.), *The German Single-Leaf Woodcut*, vol. 1, p. 93. Curiously, a song that expresses this concern was published in 1580; see Anon., *Sechs schöne newe Geistliche Lieder* (Lübeck, 1580), fol. A2r.
100. Anon., *Zwo Warhafftige Newe zeittung Vnd gründliche Geschicht* (Graz, 1593), fols A3r–A3v.
101. S. Müller, *Astronomische Beschreibung dess Cometen so zu ende dises verloffenen 1577. Jars erschienin sampt seiner bedeutung* (Augsburg, 1577), fol. A4v.
102. Licht, *Von dem schröcklichen*, fols B2v–B3r.
103. Ibid., fol. B2v.
104. *Wunderzeichen Welches zü Loschel am himmel gestanden vnnd alda ist gesehen worden der xxx. Tag Mertzen in diesem M. D. Lxxiij. Jar* (Schweinfurt: Johannes Frisch, 1573), in Strauss (ed.), *The German Single-Leaf Woodcut*, vol. 2, p. 568.
105. J. Creat, *Kurtze beschreibunge des Cometen welcher ist gesehen worden am Himmel Anno 1577. den 11. Nouembris Auch von etlichen Wunderzeichen die vorher gegangen sein* (n.p., [*c.* 1577]), fol. A1r, fol. A4v.
106. On the reactions of the Augsburg Protestants to the introduction of the Gregorian calendar, see B. A. Tlusty, *The Martial Ethic in Early Modern Germany: Civic Duty and the Right of Arms* (New York: Palgrave Macmillan, 2011), pp. 245–64.
107. *Arhafftige newe Zeitung vnd erschreckliche Wunderzeichen so sich newlichen zu Rom vnnd Paris in dem nechst vergangenen 83. Jar geshcehen* (Augsburg, 1584), fols A2r–A2v.
108. Simon Pauli's exposition 'Vom selben Cometen' is printed in David Chytraeus's tract on the 1572 nova and the 1577 comet. D. Chytraeus, *Vom Newen Stern Welcher Anno M.D.LXXII. Im Nouember erschienen: Vnd vom Cometen Welcher wir im Nouember dieses lauffenden M.D.LXXVII. Jars vnd noch jtzund sehen* (Rostock, 1577), fol. C3r.
109. B. Nischan, *Lutherans and Calvinists in the Age of Confessionalism* (Aldershot: Ashgate, 1999), essay 7: 'Lutheran Confessionalization, Preaching, and the Devil', pp. 1–20, on p. 11.
110. A. Taurer, *Hochnotwendister Bericht Von mancherley erschreckliche Wunderzeichen vnd Bußruffern* (Halle, 1592), fol. D7r.
111. J. Goodale, 'Intimidation, Intolerance, and Injury: Religious Conflict in Saxony, 1587–1592', in K. von Greyerz and K. Siebenhüner (eds), *Religion und Gewalt: Konflikte,*

Rituale, Deutungen (1500–1800) (Göttingen: Vandenhoeck & Ruprecht, 2006), pp. 195–219.

112. Nischan, *Lutherans and Calvinists*, essay 2: 'Ritual and Protestant Identity in Late Reformation Germany', pp. 142–58, on pp. 151–8.

2 Lutheran Clergy and *Wunderzeichen* Discourses

1. On Melanchthon's thoughts on astrology and astronomy, see S. Caroti, 'Melanchthon's Astrology', in P. Zambelli (ed.), *'Astrologi hallucinati': Stars and the End of the World in Luther's Time* (Berlin: Walter de Gruyter, 1986), pp. 109–21; W. Hammer, 'Melanchthon, Inspirer of the Study of Astronomy', *Popular Astronomy*, 59 (1951), pp. 308–19; C. Methuen, 'The Role of the Heavens in the Thought of Philip Melanchthon', *Journal of the History of Ideas*, 57:3 (1996), pp. 385–403; S. Kusukawa, 'Aspectio divinorum operum: Melanchthon and Astrology for Lutheran Medics', in O. P. Grell and A. Cuningham (eds), *Medicine and the Reformation* (New York: Routledge, 1993), pp. 33–56; and Methuen, *Kepler's Tübingen*, pp. 61–106.
2. P. Melanchthon, 'The Dignity of Astrology', in *Orations on Philosophy and Education*, ed. S. Kusukawa (Cambridge: Cambridge University Press, 1999), p. 124.
3. Methuen, 'The Role of the Heavens', p. 73.
4. 'An Autobiography of Martin Kemnitz', trans. A. L. Graebner, *Theological Quarterly*, 3 (1899), pp. 472–87, on p. 479, p. 481.
5. K. Hartfelder, 'Der Aberglaube Philipp Menlanchthons', *Historisches Taschenbuch*, 6 (1889), pp. 233–69, on p. 258.
6. J. Rauscher, 'Der Halleysche Komet im Jahre 1531 und die Reformatoren', *Zeitschrift fuer Kirchengeschichte*, 32 (1911), pp. 259–76, on pp. 266–7.
7. Barnes, *Prophecy and Gnosis*, p. 87.
8. L. Krentzheim, *Coniecturae: Christliche vermuttungen von künfftiger Zeit Zustandt in Kirchen vnd Regimenten* (Görlitz, 1583), fols B7r–B8v.
9. Hiob Fincel studied under Melanchthon and received a doctoral degree in philosophy at Wittenberg in 1549; Caspar Goltwurm also studied at Wittenberg in 1539–1541.
10. On the influence of Melanchthon on contemporary astronomers/astrologers, see L. Thorndike, *A History of Magic and Experimental Science*, 5 vols (New York: Columbia University Press, 1941), vol. 5, pp. 378–405; J. W. Montgomery, 'Cross, Constellation, and Crucible: Lutheran Astrology and Alchemy in the Age of the Reformation', *Ambix*, 11 (1963), pp. 65–86, on pp. 70–3; and Hammer, 'Melanchthon'. Peter Barker analyses the contributions of Melanchthon and the Wittenberg circle on the popularization of Copernican theory in 'The Role of Religion in the Lutheran Response to Copernicus', in M. J. Osler (ed.), *Rethinking the Scientific Revolution* (Cambridge: Cambridge University Press, 2000), pp. 59–88.
11. Taurer, *Hochnotwendigster Bericht*, fol. B4r.
12. On Lutheran postils, see J. M. Frymire, *The Primacy of the Postils: Catholics, Protestants, and the Dissemination of Ideas in Early Modern Germany* (Leiden: Brill, 2010); and B. Kreitzer, 'The Lutheran Sermon', in L. Taylor (ed.), *Preachers and People in the Reformations and Early Modern Period* (Leiden: Brill, 2001), pp. 35–63, on pp. 54–6.
13. In his *Table-Talk*, he left some comments about eclipses and comets that he had witnessed. See M. Luther, *Dr. Martin Luthers Sämmtliche Schriften*, ed. J. G. Walch, 25 vols (St Louis, MO: Concordia, 1887), vol. 22, p. 1554.
14. At least eight editions of this sermon were published in Wittenberg and Strasbourg.

15. M. Luther, *Sermons of Martin Luther*, ed. and trans. J. N. Lenker, 8 vols (Grand Rapids, MI: Baker Books, 1983), vol. 1, p. 66.
16. Ibid., p. 70.
17. M. Luther, *Sermons of Martin Luther: The House Postils*, ed. and trans. E. F. A. Klug, 3 vols (Grand Rapids: Baker Books, 1996), vol. 1, pp. 45–6.
18. In Saxony and some other territories, it was a legal requirement that parishes owned copies of Luther's postils. Kreitzer, 'The Lutheran Sermon', p. 54.
19. J. Wigand, *Postilla* (Frankfurt am Main, 1569), fol. B1v.
20. Ibid., fol. B2r.
21. S. Pauli, *Postilla. Theil 1* (Magdeburg, 1572), fols 18r–18v.
22. Ibid., fol. 19v.
23. Ibid., fol. 18r.
24. H. Mencel, *Postilla* (Leipzig, 1596), p. 13.
25. S. Gedike, *Postilla* (Eisleben, 1588), fol. B2r.
26. Ibid., fol. B2r.
27. G. Walther, *Außlegung der Euangelien so an den Sontagen gelesen werden. Theil. 1.* (Wittenberg, 1579), vol. 1, fol. C4v.
28. Ibid., fol. C4r.
29. Ibid., fols C4r–C4v.
30. Ibid., fol. C4v.
31. C. Vischer, *Außlegung der Euangelien: so man auff die Sontagen in der Christliche Kirchen zu handeln pfleget* (Schmalkalden, 1570), fols E2r–E2v.
32. S. Saccus, *Erklerung Vber die Sontags Euangelia vnd vber die fürnembsten Fest durchs gantze Jahr* (Magdeburg, 1592), p. 43.
33. M. Mirus, *Postilla* (Jena, 1605), fol. B5r.
34. M. Chemnitz, *Postilla* (Mageburg, 1594), p. 532.
35. A. Hunnius, *Postilla* (Wittenberg, 1592), fol. D4v.
36. J. Spangenberg, *Postilla Vom Aduent bis auff Ostern: Für die jungen Christien Knaben vnd Meidlein in Fragstücke verfasset* (Erfurt, 1564), fol. B4v.
37. For example, Johann Brenz and Simon Musaeus do not show much interest in *Wunderzeichen* in their comments on Christ's prophecy, and I could not find any discourse about celestial phenomena in Tilemann Hesshus's postil.
38. A. Schomer, *Sieben Bußpredigten Von den zukünfftigern schrecklichen straffen großen verenderung vnd grewlichen verhanden Trübsalen so für dem Jüngsten tage fürher gehen vnd sonderlich vber Deutschland kommen werden* (Lemgo, 1583), fol. D3r.
39. Ibid., fol. D3v.
40. B. Faber, *Allerley Christliche nötige vnd nützliche vnterrichtungen von den Ketzen Hendeln der Welt* (Eisleben, 1565), fol. B4r.
41. C. Spangenberg, *Ernste vnd hochnöttige Busspredigt An gantzes Deutschland* (Eisleben, 1569), fol. N1r; W. Peristerus, *Eine Trewhertzige Busspredigt Vber den Straffspruch Christi aus dem Euangelio des 21. Sontags nach Trinitatis* (Berlin, 1580), fol. D1r.
42. Schütz, *Wunderzeichen*, fol. B8v.
43. A. Musculus, *Vom Mensch vnd Kedar, vom Gog vnd Magog* (Frankfurt an der Oder, 1577), fol. C3r.
44. J. Heerbrand, *Ein Predig Von dem erschrocklichen Wunderzeichen am Himmel dem newen Cometen oder Pfawenschwantz Gehalten zu Tübingen den 24. Sontag nach Trinitatis wolcher ist der 17. Wintermonats* (Tübingen, 1577), fol. A4v.

45. A. Musculus, *Weissagung D. Martini Lutheri der Deustchen Apostel vnd Prophet von dem grossen zunahenden vnglück über Deutschland durch den Gog Magog vnd Mesch* (Frankfurt an der Oder, 1568), fols A2v–A3r, fol. E5v.
46. C. Irenaeus, *Abdruck: Eines schrecklichen Zornzeichens Sampt Christlicher vnd nötiger Erinnerung* (Eisleben, 1564), fols A4r–B2r.
47. H. Opitz, *Von dem erschrecklichen vnnd grossen Fewrigen Zeichen welches am Himel am tag der vnschüldigen Kindtlein im Jar nach der geburt Christi M.D.LXI. An vielen orten vnd Stedten ist gesehen worden* (Nuremberg, 1561), fols A1v–A2r.
48. Gernhard, *Vom Jüngste Tage*, fols B3v–B4r.
49. J. Coler, *Notwendige Erinnerung auff dz schreckliche Fewerzeichen So den 10. Septembris deß jetzlauffenden 1580. Jahrs am Himmels gesehen worden* (Berlin, 1581), fols B3r–B3v.
50. Peristerus, *Bußpredigt*, fol. D1r.
51. M. Bapst, *Warhafftige beschreibung Des Erschrecklichen Blut vnd Fewerzeichens welches den 6. Martij dieses vnd ablauffenden 1582. Jahs am Himmel gesehen worden* (Freiberg, 1582), fol. A2v.
52. Spangenberg, *Manfeldische Chronica*, fol. 480v.
53. J. Gölitz, *Newe Zeitung: Ein Wunderbarlich Gesicht am Himmel so durch den Pfarrhen vnnd ander Personen des Dorffs Aschre inn der Löblichen Herrschafft Thoma gelegen gesehen ist worden* (n.p., 1560).
54. Dixon, 'Popular Astrology', p. 409.
55. See G. Strauss, 'Success and Failure in the German Reformation', *Past and Present*, 67 (1975), pp. 30–63; G. Strauss, *Luther's House of Learning: Indoctrination of the Young in the German Reformation* (Baltimore, MD: Johns Hopkins University Press, 1978), chs 12, 13, 14; and G. Parker, 'Success and Failure during the First Century of the Reformation', *Past and Present*, 136 (1992), pp. 43–82. There are some arguments as to whether visitation records are reliable sources to measure people's spirituality. For example, see E. M. Kern, 'The "Universal" and the "Local" in Episcopal Visitations', in M. Reinhart (ed.), *Infinite Boundaries: Order, Disorder, and Reorder in Early Modern German Culture* (Kirksville, MO: Sixteenth Century Journal Publishers, 1998), pp. 35–54. Recent studies such as Kern's emphasize regional differences.
56. Opitz, *Von dem erschrecklichen*, fols A2v–A3r.
57. Schomer, *Sieben Bußpredigten*, fol. D3r.
58. N. Erben, *Zur Buss vermanende Comet: Anno M.D.LXXII. Den XIIII. Maij den abendt vmb VIII. Vhr erschienen* (Erfurt, 1582), fols A3r–A3v.
59. Musculus, *Weissagung*, fols D1v–D2r.
60. Ibid., fol. D3v.
61. Ibid., fol. V6r.
62. A. Musculus, *Gebet Vmb abwendung woluerdienter zunahender Straff vber Deutschland* (Erfurt, 1569), fol. D3v.
63. P. Eber, *Beschreibung des schrecklichen zeichens so am 13. tag Martij fast die gantze nacht vber zu Wittenberg vnd an viel andern orten ist gesehen worden* (Wittenberg, 1562), fol. B4v.
64. Ibid., fol. C1r, fols B1v–B2r, fols B2v–B3r.
65. Ibid., fols C1r–C1v.
66. Ibid., fol. C2r.
67. Bapst, *Warhafftige beschreibung*, fols B1r–B1v; Victorius, *Xasmatodogia*, fols C2v–C3r; Cuno, *Erinnerung*, fols E1v–E2r.
68. A. Celichius, *Theologische erinnerung von dem newen Cometen* (Magedeburg, 1578), fols B2v–B3r.

69. For example, J. Scheitlich, *Warnung für der grossen erschrecklichen vnd nun fast zu nahender straffen Gottes vber das gantze Deustchland* (Eisleben, [*c*. 1565]); Spangenberg, *Ernste vnd hochnöttige Busspredigt*; and D. Meder, *Zehen Christliche Busspredigten vber Die Weissagung Christi dess grossen Propheten vom Ende der Welt vnd Jüngste Tage* (Frankfurt am Main, 1581).
70. Taurer, *Hochnotwendigster Bericht*, fols G3v–G4r.
71. Strauss, *Luther's House of Learning*, pp. 212–13.
72. J. Wigand, *Vom Straffampt der Sünden* (Ursel, 1564), fol. B4v.
73. *Dieses erschreckliche Zeichen ist am Himel gegen auffgang der Sonnen zu Marburg in Hessen den xvij. Octobris dieses 1570. Jhars gesehen worden* (Erfurt, 1571); in Harms et al. (eds), *Herzog August Bibliothek I*, p. 377 (plate 184).
74. A. Praetorius, *Selige Erinnerung Von dem erschrecklcihen Fewerenden Cometstern am Himel* (Erfurt, 1580), fols A2r–A2v.
75. Bapst, *Warhafftige beschreibung*, fol. B2r.
76. Peristerus, *Bußpredigt*, fol. B4r.
77. Ibid., fol. E1v.
78. Pfeffinger, *Christliche*, fol. B3v.
79. Ibid., fol. B4v.
80. Ibid., fol. C1v.
81. *Sermons of Martin Luther*, ed. and trans. Lenker, p. 80, p. 77.
82. Pfeffinger, *Christliche*, fol. C1r.
83. Cuno, *Erinnerung*, fol. F3r.
84. Suevus, *Erbauungsschriften*, fol. 170v.
85. Ibid., fols 143v–144r.
86. J. A. Conner, *Kepler's Witch: An Astronomer's Discovery of Cosmic Order amid Religious War, Political Intrigue, and the Heresy Trial of his Mother* (San Francisco, CA: HarperSanFrancisco, 2004), p. 26. The illustration printed on a Nuremberg broadsheet of the 1577 comet depicts a mother showing the comet to a little boy, a scene that reminds us of Kepler's episode. *Verzaichnuß des Cometen so im Nouemb: in disem 77. jar Zum ersten mal gesehen worden* (Nuremberg: Georg Mack, 1577); in Harms and Schilling (eds), *Wickiana II*, p. 171 (plate 86).
87. *Ein grausamb vnd erschröcklich wunderzeichen so am 28. tag Decembris im L. X. Jar zu Eckelsheym ein Meyl wegs von Forchheym geschehen ist* (Nuremberg, 1560); in Harms and Schilling (eds), *Wickiana I*, p. 219 (plate 109).
88. Opitz, *Von dem erschrecklichen*, fol. A2v. It seems that he received this request while the fiery beams were still flying in the sky because he says 'today, the Day of the Holy Innocents' in his account (fol. A1v).
89. Peristerus, *Trostreiche Predigt*, fol. E1v.
90. R. Scribner, 'Anticlericalism and the German Reformation', in *Popular Culture and Popular Movements in Reformation Germany* (London: Hambledon Press, 1987), pp. 243–56, on pp. 254–6.
91. G. Strauss, 'Local Anticlericalism in Reformation Germany', in P. A. Dykema and H. A. Oberman (eds), *Anticlericalism in Late Medieval and Early Modern Europe* (Leiden: Brill, 1993), pp. 625–37.
92. S. C. Karant-Nunn, 'Neoclericalism and Anticlericalism in Saxony, 1555–1675', *Journal of Interdisciplinary History*, 29:4 (1994), pp. 615–37.

93. S. Pauli, 'Vom selben Cometen Erinnerung warnung vnd vermanu[n]g'. This piece is printed in Chytraeus's pamphlet about the nova and the comet, *Vom Newen Stern Welcher Anno M.D.LXXII*, fol. C4r.
94. Caesius, *Chronick*, fol. T1r.
95. A. Praetorius, *Schade Wacht: Nothwendige Warnung vnd Vermanung an alle busfertige glaubige Christen getahn von dem erschrecklichen Comet vnd Zornzeichen der dieses 77. Jar am Himel gesehen wird* (Erfurt, 1577), fol. A2r.
96. Ibid.
97. N. Selnecker, *Ein Christlich Gebet in jetzigem eleneden zustand darin Gott selbs vns seine fewrige Rute vnd seinenn gerechten Zorn am Himmel zeiget vnd vns mit allerley drawungen vnd züchtigungen heimsuchet zuthun vnd sich damit für Gott demütiglich anzugehen vmb Gnad zu bitten* (Leipzig, 1578), fol. A3r.
98. Ibid.
99. Eber, *Beschreibung*, fol. C1v.
100. H. Weller, *Wunderliche vnd vnerhorte Geschicht die sich zu Freybergl vnnd inn derselben gegent inn Meissen zugetragen hat den xiij. Augusti / Jm M.D.Lix* (Magdeburg, 1559), fol. A2v.
101. The Stendal pastor Daniel Schaller uses this prase ('trewer Eckhart') when he discusses God's mercy shown through *Wunderzeichen* in D. Schaller, *Theologischer Heroldt* (Magdeburg, 1604), fol. N3r.

3 'An Eagle Hurting Himself': Flacius's Tract against the Interim

1. R. Kolb and J. Nestingen (eds), *Sources and Contexts of the Book of Concord* (Minneapolis, MN: Fortress Press, 2001), pp. 150–4.
2. For example, see C. L. Manschreck, 'The Role of Melanchthon in the Adiaphora Controversy', *Archiv für Reformationsgeschichte*, 48 (1957), pp. 165–82.
3. T. J. Wengert, '"Not by Nature *Philoneikos*": Philip Melanchthon's Initial Reactions to the Augsburg Interim', in *Philip Melanchthon, Speaker of the Reformation* (Farnham: Ashgate, 2010), Essay XIII, pp. 33–49, on p. 46.
4. R. Kolb, 'Philipp Melanchthon: Reformer and Theologian', *Concordia Journal*, 23:4 (1997), pp. 309–16, on p. 314.
5. O. K. Olson, *Matthias Flacius and the Survival of Luther's Reform* (Wiesbaden: Harrassowitz, 2002).
6. This is the voice of Nikolaus von Amsdorf, one of the leaders of the Anti-Interim campaign. R. Kolb, 'Nikolaus von Amsdorf, Knight of God and Exile of Christ: Piety and Polemic in the Wake of Luther' (PhD dissertation, University of Wisconsin, 1973), p. 185.
7. Originally this word was a derogatory term against Magdeburg theologians used by Wittenberg theologians. O. K. Olson, 'Theology of Revolution: Magdeburg, 1550–1551', *Sixteenth Century Journal*, 3:1 (1972), pp. 56–79, on p. 70.
8. Details of the resistance theory of Magdeburg are discussed in ibid.; and D. M. Whitford, *Tyranny and Resistance: The Magdeburg Confession and the Lutheran Tradition* (St Louis, MO: Concordia, 2001).
9. N. Rein, *The Chancery of God: Protestant Print, Polemic and Propaganda against the Empire, Magdeburg 1546–1551* (Aldershot: Ashgate, 2008), p. 16; T. Kaufmann, '"Our Lord God's Chancery": Magdeburg and its Fight against the Interim', *Church History*, 73:3 (2004), pp. 566–82, on p. 572.
10. Kaufmann, 'God's Chancery', p. 573.

11. Rein, *The Chancery of God*, pp. 170–2.
12. Nikolaus von Amsdorf describes five signs of the imminent end of the world in his book, which criticizes the Interims. They are: mass apostasy; the revelation of Antichrist; the decline of the Roman Empire; the encroachment of human traditions upon the Gospel; and people's worship of the 'beast'. N. von Amsdorf, *Fünff furnemliche vnd gewisse Zeichen aus heiliger göttlicher Schrifft so Kurtz vor dem Jungsten tag geschehen sollen* (Jena, 1554).
13. Rein, *The Chancery of God*, p. 146.
14. According to Kaufmann, Flacius wrote or was involved in 40 per cent of the publications in Magdeburg during 1549–50. Kaufmann, 'God's Chancery', p. 576.
15. M. Flacius Illyricus, *Eine Weissagung vnd ein schöner Herrlicher trost für alle hochbetrübte frome Christliche hertzen zu diser jtzigen tru[b]seligen zeit: Aus dem XIIII. Cap. Der offenbarung Johannis* (Magdeburg, 1548), fol. A2r, fols A2r–A2v, fol. B1r.
16. Ibid., fol. B1v. The word 'Interim' is singular, but it must signify both the Augsburg Interim and the Leipzig Interim.
17. Ibid., fols B2r, B3r, B4r.
18. M. Flacius Illyricus, *Kleriche beweisung das alle die jenige welche die schrifften widder das INTERIM vnd Mitteldingfeil zuhaben vn[d] zu lesen verbieren* (Magdeburg, 1550), fols A3r–A3v, fol. A7v, fols A8r–A8v.
19. R. W. Oettinger, *Music as Propaganda in the German Reformation* (Aldershot: Ashgate, 2001), p. 188.
20. P. Melanchthon and M. Flacius Illyricus, *Ein grausam Meerwunder den Bapst bedeutende zu Rom gefunden* (Mageburg, 1550).
21. In the illustration, the wagon is drawn by two horses, and the driver is riding one of them.
22. N. Medler and M. Flacius Illyricus, *Zeychen am himmel bey Braunschwig newlich gesehen: durch den superatendentem zu Braunschwig geschriben Mit einer vorrede* (Magdeburg, 1549).
23. Ibid., fols A2r–A2v.
24. Ibid., fol. A2v, fol. A3r, fol. A3v.
25. The word 'Mameluke', which meant the slave soldiers of Muslim rulers, often appears in the Gnesio-Lutherans' attack on their theological opponents. In general the word signifies apostates, probably because among those slave soldiers there were some ex-Christians who renounced their faith.
26. Medler and Flacius, *Zeychen*, fol. A4r.
27. A group of Magdeburg citizens tried to kidnap a Jew named Michael, a subject of Elector Joachim of Brandenburg, and the report of their execution was widely circulated. Flacius did not explain the details in his tract, probably because he believed that people knew well about this case. Rein, *The Chancery of God*, pp. 154–5.
28. Medler and Flacius, *Zeychen*, fol. A4r.
29. Ibid., fol. A4v.
30. Ibid.
31. Ibid., fol. B1r. Obviously two darker moons were *Nebensonnen*.
32. Ibid., fol. B1r, fols B1r–B1v, fol. B1v.
33. Ibid., fol. B2r, fol. B2v.
34. Ibid., fol. B3r.
35. T. Kaufmann, *Ende der Reformation: Magdeburgs 'Herrgottes Kanzlei' (1548–1551/2)* (Tübingen: Mohr Siebeck, 2003), p. 512.
36. According to Kaufmann, Kempff published only five titles in 1547–52, while Lotter printed 190 titles and Roeddinger 145; ibid., p. 562. The title of the song is 'Des Interms,

vnd Intermisten warhafftige abgemalte figure', illustrated by the monogrammist 'BP'. Oettinger, *Music as Propaganda*, p. 145.

37. S. L. Jansen, *Political Protest and Prophecy under Henry VIII* (Woodbridge: Boydell Press, 1991). See also R. Taylor, *The Political Prophecy in England* (New York: Columbia University Press, 1911).
38. J. Carion, *Außlegung der verborgenen Weissagung D. Johan. Carionis, von veranderung und zuvfelligem glück der hösten Potetaten des Römischen Reichs* (Strasbourg, 1548), fols A2r–A4r.
39. Medler and Flacius, *Zeychen*, fol. B1r.
40. For example, a broadsheet (1562) about the appearance of various apparitions in the sky of Hamburg reports that a big man in Turkish dress and a camel (*Camelthier*) were also seen. Strauss (ed.), *The German Single-Leaf Woodcut*, vol. 1, p. 194.
41. Goltwurm, *Wunderwerck*, fol. Oo1v. Goldwurm's narrative of the Braunschweig case is much abridged. He does not explain the circumstance of the apparitions and mentions only John Frederick's picture, the images of Adam and Eve, Christ on the cross between the two thieves, the apostles' gathering, and a man with a sword and a weeping maid. Three moons, a lion, an eagle, a camel, castles and people were omitted.
42. Soergel, *Miracles*, p. 95.
43. Kaufmann, 'God's Chancery', pp. 574–5.
44. Rein, *The Chancery of God*, p. 168.
45. Anon., *Etliche gesichte sozu Holtzhausen vnther Wasserburg im Lande During gelegen am Donnerstag nach Trinitatis Und zu Pretin den xx. Junij. Diese xlviij Jars am Himmel von glaubwirdigen Leuten seint gesehen worden* (n.p., 1548). Thomas Kaufmann mentions the two cases in his *Ende der Reformation*, p. 93.
46. Anon., *Etliche gesichte*, fols A2r–A2v, fol. A2v, fol. A3r, fol. A3v.
47. Fincel, *Wunderzeichen*, fols P7v–P8v; Goltwurm, *Wunderwerck*, fol. Ji4v.
48. B. Monner, *Bedencken vonn dem Kriege der Anno rc. Sechs siben vnd viertzig im land zü Meissen vnd Sachsen gefürt ist so für er zühalten sey gestalt* (Basel, 1557), fol. F1v.
49. Kaufmann, *Ende der Reformation*, p. 93.
50. *Ein wunderbarliche vnd warhafftige geschicht so von meniglich gesehen vnd gehört ist owrden Jn der Churfürstlichen Statt Wittemberg in Sachsen gelegen den Achtzehenden tag Herbstmonat im Jar M. D. XLVII* (Strassburg, 1547); in Harms and Schilling (eds), *Wickiana I*, p. 79 (plate 38).
51. Olson, *Matthias Flacius*, p. 88.
52. P. E. Starenko, 'In Luther's Wake: Duke John Frederick II of Saxony, Angelic Prophecy, and the Gotha Rebellion of 1567' (PhD dissertation, University of California Berkeley, 2002), p. 95.
53. *Des Gefangnen Christlichen Churfürsten, rechter Titel so ihn itziger Zeit, von allen Gottseligen Waren Christen, billich gegeben wird zu Latein vnd Deutsch in ein Lied verfasset* (Magdeburg: Michael Lotter, n.d.), cited in Olson, *Matthias Flacius*, p. 93.
54. Rein, *The Chancery of God*, p. 75. This text was preserved only in a manuscript copy. It reminds us of the sixteenth-century text that describes the ordeal of Luther at Worms with the Passion of Christ.
55. *Vier schöner lieder das erste des Churfürsten zu Sachssen: im Thon ein newes lied wir haben an.* (n.p., [*c.* 1549]).
56. Rein, *The Chancery of God*, p. 47.

57. *Unterscheid zwischen der waren Religion Christi vnd falschen Abgöttishcen Lehr des Antichrists* (Magdeburg, n.d.). The whole illustration and the text are in Kaufman, *Ende der Reformation*, pp. 586–7.
58. Carl C. Christensen analyses the background and iconography of this print in *Princes and Propaganda: Electoral Saxon Art of the Reformation* (Kirksville, MO: Sixteenth Century Journal Publishers, 1992), pp. 79–88.
59. Monner's argument is analysed in R. Kolb, 'The Legal Case for Martyrdom: Basilius Monner on Johann Friedrich the Elder and the Smalcald War', in I. Dingel, V. Leppin and C. Strohm (eds), *Reformation und Recht* (Gütersloh: Chr. Kaiser, 2002), pp. 145–60.
60. Monner, *Bedencken*, fols F1v–F2r.
61. Ibid., fols F3v–F4r.
62. Kolb, 'Legal Case', p. 147.
63. Fincel, *Wunderzeichen*, fol. Q8r.
64. 'Talia simulacra quae causae in aere pignant, saepe quaesitum est. Ego pleraque divinitus ostendi per bonos angelos existimo; interdum et diabolorum praestigas esse consentaneum est. Multa autem significantia ostendunt eventus, ac aliquot locis iam in coelo conspecta imago Cameli haud dubie denuntiat vastationes, quas in Germania Turcos facturos esse metuendum est. Intervallum vero aliquod significatum videtur specie creationis Evae, quae significant, mediocrem Ecclesiae statum. Postea sequentur bella Turccia, et exercebitur magna crudelitas adversus Ecclesiam Dei. Haec ita futura esse et *Hiltenius* Isenaci praedixit ante annos fere centum'. P. Melanchthon, *Corpus Reformatorum. Philippi Melanthonis opera quae supersunt omnia*, ed. K. Gottlieb Bretschneider, 28 vols (Halle: C. A. Schwetschke, 1840), vol. 7, p. 653.
65. A. S. Francisco, *Martin Luther and Islam: A Study in Sixteenth-Century Polemics and Apologetics* (Leiden: Brill, 2007), pp. 22–3.
66. Krentzheim, *Coniecturae*, fols b7r–b7v.
67. Ibid., fol. b7v, fol. b8r, fol. B8v, fols b8r–b8v, fols b8v–c1r, fol. c1r.
68. Barnes, *Prophecy and Gnosis*, p. 269.
69. M. Flacius Illyricus, *Gründliche Verlegung des Langen Comments der Adiaphoristen oder der verzelung jrer handlungen: Zu gründlicher erforschung der warheit in dieser sache sehr nützlich zu lessen* (Jena, 1560), fol. Gg3r.
70. Ibid., fol. Gg3v.
71. Ibid., fol. Gg3r.
72. For example, Robert Kolb suggests that one of the reasons why Melanchthon submitted to Duke Moritz's demand is that the stars showed the prospect of Charles's death: 'In addition, he believed that the stars foretold an early death for Charles V; buying a little time through compromise would be worthwhile since a new emperor would soon mount the throne'. See Kolb's introduction to J. Andreae, *Andreae and the Formula of Concord: Six Sermons of the Way to Lutheran Unity*, ed. and trans. R. Kolb (St Louis, MO: Concordia, 1977), p. 23.
73. J. Schofield, *Philip Melanchthon and the English Reformation* (Aldershot: Ashgate, 2006), p. 158.
74. Melanchthon, *Corpus Reformatorum*, vol. 7, p. 758.
75. Ibid., p. 759.
76. Ibid. This *Wunderzeichen* case is reported in a broadsheet published by Steffan Hamer of Nuremburg (1550).
77. M. Wriedt, 'Between Angst and Confidence: Melanchthon as a Man of the Sixteenth Century', *Concordia Journal*, 23:4 (1997), pp. 277–94, on p. 285.

78. Fincel, *Wunderzeichen*, fol. P3v.
79. A. Duke, 'A Legend in the Making: News of the "Spanish Inquisition" of the Low Countries in German Evangelical Pamphlets, 1546–1550', *Nederlandsch archief voor kerkgeschiedenis*, 22 (1997), pp. 125–44, on p. 120.
80. Fincel, *Wunderzeichen*, fol. P4r. This story became popular in Protestant lore and is repeatedly quoted in Lutheran discourses on natural disasters in the early seventeenth century.
81. M. Flacius Illyricus, *Ordenung vnd Mandat Keiser Caroli V. vernewert im April Anno 1550* (Magdeburg, 1550), fol. A1v.
82. Ibid., fol. A3r. This number is obviously exaggerated, but it was true that several hundred Protestants had been executed in the Netherlands by 1550 by order of the emperor.

4 Irenaeus against 'Spiritual Wolves': Polemical Use of *Wunderzeichen*, I

1. The first controversy, the Majoristic Controversy, sprang from Georg Major's thesis 'good works are necessary for salvation' (1552). Soon after the Synergist Controversy over the role of human will in the process of salvation broke out (1555). Later this controversy led to another one, the Flacian Controversy over the power of original sin (1560). In addition, Andreas Osiander triggered another controversy in the 1550s by claiming that one was justified by indwelling of Christ with his divine righteousness. During the 1570s, the Gnesio-Lutheran theologians expressed concern that some theologians of Wittenberg had drifted away from Luther's 'real presence' understanding of the Lord's Supper and were promoting a spiritualizing view of Christ's presence in the Communion. Gnesio-Lutherans attacked such tendencies as crypto-Calvinism. On the general history of Lutheran theological controversies, the following studies are useful: C. P. Arand, J. A. Nestingen and R. Kolb, *The Lutheran Confession: History and Theology of the Book of Concord* (Minneapolis, MN: Fortress Press, 2012); I. Dingel, 'The Culture of Conflict in the Controversies Leading to the Formula of Concord (1548–1580)', in R. Kolb (ed.), *Lutheran Ecclesiastical Culture* (Leiden: Brille, 2008), pp. 15–64; E. F. Klug, *Getting into the Formula of Concord: A History and Digest of the Formula* (St Louis, MO: Concordia, 1977); and Kolb, 'Nikolaus von Amsdorf'.
2. R. Kolb, 'The Flacian Rejection of the Concordia, Prophetic Style and Action on the German Late Reformation', *Archiv für Reformationsgeschichte*, 73 (1982), pp. 196–216.
3. R. Christman, *Doctrinal Controversy and Lay Religiosity in Late Reformation Germany: The Case of Mansfeld* (Leiden: Brill, 2012); see also R. Christman, 'Heretics in Luther's Homeland: The Controversy over Original Sin in Late Sixteenth-Century Mansfeld' (PhD dissertation, University of Arizona, 2004).
4. I. Dingel, *Concordia Controversa: Die öffentlichen Diskussionen um das lutherische Konkordienwerck am Ende des 16. Jahrhunderts* (Gütersloh: Gütersloher Verlagshaus, 1996), esp. pp. 499–514.
5. Barnes, *Prophecy and Gnosis*, p. 90.
6. Schenda, 'Prodigensammlungen', pp. 657–63.
7. R. Schenda, 'Irenaeus, Christoph', in *Enzyklopädie des Märchens*, 14 vols (Berlin: Walter de Gruyter, 1991) vol. 7, pp. 270–4.
8. Leppin, *Antichrist und Jüngster Tag*, pp. 109–12. Leppin also mentions Irenaeus's works in some footnotes.

9. Dingel, *Concordia Controversa*, pp. 521–2.
10. Soergel, *Miracles*, pp. 124–52.
11. Ibid., p. 131.
12. Ibid., pp. 137–8.
13. The following basic biographical information about Irenaeus is mainly based on: G. Bossert, 'Irenaeus, Christoph', in P. Schaff (ed.), *The New Schaff-Herzog Encyclopedia of Religious Knowledge*, 13 vols (Grand Rapids, MI: Baker Books, 1953), vol. 6, p. 31; Christman, *Doctrinal Controversy*, pp. 501–2; G. Frank, 'Irenaeus, Christoph', in *Allgemeine Deutsche Biographie*, 56 vols (Leipzig: Duncker & Humblot, 1881), vol. 14, p. 582; H. Junghans, 'Irenäus, Christoph', in *Neue Deutsche Biographie*, 25 vols (Berlin: Duncker & Humblot, 1974), vol. 10, p. 178; and K. Lohrmann, 'Irenäus, Christoph', in *Biographisch-Bibliographisches Kirchenlexikon*, 14 vols (Hamm: Traugott Bautz, 1990), vol. 2, pp. 1327–8.
14. Bossert, 'Irenaeus, Christoph'.
15. C. Irenaeus, *Symbolvm Apostolicvm: Das ist Die Artickel vnsers Christlichen Glaubens ausgelegt. Mit Vermeldung vnd widerlegung allerley Jrthumb vnd Ketzerien beide alt vnd new so darwider entstanden* (Eisleben, 1562–3), fol. B3v.
16. Spangenberg, *Mansfeldische Chronica*, fol. 487v.
17. Kolb, 'Flacian Rejection', pp. 199–200.
18. C. Irenaeus, *Warnung vnd vrsachen: Das man nicht in eine Amnistiam vnd Stilschweigen der Jrthumen vnd Corruptelm Noch in den Orden der newen Jacobsbrüder willigen soll* (n.p., 1569), fol. D2v.
19. Kolb, 'Flacian Rejection', p. 201.
20. Andreae, *Andreae and the Formula of Concord*, ed. Kolb, p. 44.
21. *Bekentniß von diesen zweyen Proposition oder Reden Peccatum originis est subtantia vnd Peccatum originis est accidens* (n.p., 1572).
22. J. Wigand, *Vom Flickwerck M. Jrenaej* (Jena, 1572).
23. Christman, *Doctrinal Controversy*, p. 12.
24. Bossert, 'Irenaeus, Christoph'; Lohrmann, 'Irenäus, Christoph'.
25. M. P. Fleischer, 'The Reception in Silesia' in L. W. Spitz and W. Lohff (eds), *Discord, Dialogue, and Concord: Studies in the Lutheran Reformation's Formula of Concord* (Philadelphia, PA: Fortress Press, 1977), pp. 119–35, on p. 122.
26. Only Bossert (in 'Irenaeus, Christoph') mentions Irenaeus's sojourn in these regions.
27. Junghans, 'Irenäus, Christoph'.
28. In his *Gnadenspiegel*, Irenaeus states that his first wife's epitaph is in Buchenbach. C. Irenaeus, *Euangelischer Vnd tröstlicher Gnadenspiegel Wieder den Schrecklichen Zornspiegel des Gesetzes* (Ursel, 1593), fol. Pp3r. In the same year, Irenaeus married a woman named Veronika, but her family name and background are not known. See Junghans, 'Irenäus, Christoph'.
29. Soergel, *Miracles*, p. 146.
30. *Christlich Bekandtnus Einhelliger Consens Bedencken vnd Rathschlag* (n.p., 1587).
31. The full English translation of this text is printed in Andreae, *Andreae and the Formula of Concord*, ed. Kolb, pp. 58–120.
32. The Torgau Book is based on both the Swabian-Saxon Concord and the Maulbronn Formula, the confession compiled by two Württemberg theologians. On the history of the formulation of the Formula of Concord, see T. R. Jungkuntz, *Formulators of the Formula of Concord: Four Architects of Lutheran Unity* (St Louis, MO: Concordia, 1977); and R. Kolb, 'Parties, Princes, Pastors, and Peace: The Formulation of Concord, 1577', *Academy: Lutherans in Profession*, 34 (1977), pp. 2–13.

33. L. W. Spitz, 'The Formula of Concord Then and Now', in L. W. Spitz and W. Lohff (eds), *Discord, Dialogue, and Concord: Studies in the Lutheran Reformation's Formula of Concord* (Philadelphia, PA: Fortress Press, 1977), pp. 1–12, on p. 7.
34. These works include *Mercklich Partickel: Der langst gesuchten Formel Concordien eynigkeit vnd vergleichung D. Jacobs Andreae in Religions sachen* (1580); *Examen des ersten Artickels vnd des Wirbel Geists, Im newen Concordien Buch* (1581); *Vordrab Vnd Wirbel Geist, So sich im newen Concordien Buch sehen lest* (1581); *Von dem Newen Dogmata der Todten Erbsünder, vnnd die Seelig im Herren verstorbenen Leichnamschender* (1583); and *Wächterhörnlein: Vnd trewe warnung für dem Gottslesterlichen vnd hochschädlichen Jrthumb der Concordisten oder Accidens-schwermer* (1583).
35. M. Chemnitz, N. Selnecker and T. Kirchner, *Refutatio Irenaej: Grundlicher Bericht auff das Examen M. Christophori Irenej, so er Anno 1581. wider den ersten Artickel deß Christilichen ConcordiBuchs* (Heidelberg, 1583).
36. Irenaeus, *Symbolvm Apostolicvm*, fol.)(3r.
37. Ibid., fol.)(6v.
38. Ibid., fol.)(8v. Irenaeus borrowed this paragraph when he wrote *Warnung vnd Vrsach* in 1569.
39. Irenaeus, *Symbolvm Apostolicvm*, fols O2r–O2v.
40. Ibid., fol. P3r.
41. Ibid., fols Q3v–Q4r. M. Luther, *Lectures on Genesis*, in *Luther's Works*, 55 vols (St Louis, MO: Concordia, 1957–86; 1958), vol. 1, p. 145.
42. Irenaeus, *Symbolvm Apostolicvm*, fol. R3r.
43. Ibid., fol. O4r. He also adds one more role of stars: that they enable us to make a calendar and count and calculate days. Ibid., fol. O4v.
44. Ibid., fol. P2r.
45. Ibid., fol. P2r.
46. J. Stoltz and J. Aurifaber, *Kurtze Verlegung der vnchristlichen Practica Magistri Johannis Hebenstreits auff das jar 1554 zu Erffurd ausgangen* (Jena, 1554).
47. Irenaeus, *Symbolvm Apostolicvm*, fol. V2v.
48. We know that something real appeared in the sky of Saxony during the night because there are several sources that record this incident. Spangenberg did not witness the phenomenon because he was visiting other places at the time, but his description of the incident in his Mansfeld chronicle is probably based on what he heard from his friends in Eisleben – maybe from Irenaeus himself. Indeed, Spangenberg's description generally agrees with Irenaeus's narrative. Spangenberg, *Mansfeldische Chronica*, fols 483v–484r. At least one broadsheet about this *Wunderzeichen* was published, though the content is different from Irenaeus's narrative. According to this broadsheet, many people of Leipzig saw 'two columns' ('zwo Seulen') of fiery flames burning in the clouds. *Ein Erschrocklich Gesicht auff den XIX Feburuarij dises 1564. Jars zu Leipzig von vilen Namhafften Personen ist am Hellen Himmel gesehen worden* (Augsburg, 1564), in Strauss (ed.), *The German Single-Leaf Woodcut*, vol. 3, p. 894.
49. Irenaeus, *Abdruck*, fol. A4v, fol. B1r, fol. B1v, fol. B2r.
50. Ibid., fols C1r–C2r.
51. Ibid., fol. C2v.
52. Ibid., fol. D1v. It is obvious that the source of Irenaeus's narrative of these *Wunderzeichen* cases was Hiob Finzel's *Wunderzeichen*.

53. Irenaeus, *Abdruck*, fols D1v–D2v. Though they did occur in his century, Irenaeus also includes stories of a sword in the sky before the rise of Mahomet and of celestial signs in Italy before the rise of the popes.
54. Ibid., fol. D2v.
55. Ibid., fol. D3r. This section does not have a number, but since it is between 4 and 6, it must be 5.
56. Ibid., fol. D4v.
57. Ibid., fol. F2v.
58. Ibid., fol. G1r.
59. Ibid., fol. H3r.
60. Irenaeus lists the devils controlling German people, such as the 'lying- or heresy-devil', 'curse-devil', 'anger- and murder-devil', 'heavy-drinking-devil' and 'usury- devil' (ibid., fol. H3b) This list of devils shows Irenaeus's familiarity with *Teufelsbucher* (devil-books), a literary genre popular in Lutheran Germany between the 1550s and the 1600s. In these books, it is depicted that each devil takes charge of a specific vice and works to tempt people to commit these sins. This literary image offered writers a convenient device to attack specific vices among people. Several of them, including C. Spangenberg's *Hunting-Devil* (Jagteuffel, 1560), were published from Urban Gaubish's press in Eisleben in the 1560s. See K. L. Roos, *The Devil in 16th Century German Literature: The Teufelsbücher* (Bern: Herbert Lang, 1972).
61. Irenaeus, *Abdruck*, fol. H3v.
62. R. Kolb, *Martin Luther as Prophet, Teacher, and Hero: Images of the Reformer, 1520–1620* (Grand Rapids, MI: Baker Books, 1999), pp. 178–83.
63. Irenaeus, *Abdruck*, fol. H4r.
64. Ibid., fols J1r–J2r.
65. Ibid., fol. K2r, fol. K1v, fol. K2r, fols K2r–K2v, fol. K2v, fols K2v–K3r, fol. K3v.
66. C. Irenaeus, *Wasser-Spiegel* (Eisleben, 1566), fol. B1r, fols B2v–B3v.
67. Ibid., fols C3v–C7v.
68. Ibid., fol. D1r, fol. D2r.
69. Ibid., fols H3r–H3v, fols H4r–H5r, fols H5v–H6r.
70. Ibid., fol. O4r.
71. Ibid., fols P4v–P5v.
72. Ibid., fols S3r–S3v, fol. S3v, fol. S4v.
73. Ibid., fols S8v–T1r, fols T2r–T2v.
74. Ibid., fols T5r–5v, fols T7r–T8v, fol. T8v, fols V2v–V3r.
75. Ibid., fol. Y2r.
76. The Gospel has been spread through whole world; sects and heresy are flooding the Church; people indulge in sins (ibid., fol. k6r); they laugh at preaching and *Zornzeichen* and do not fear God's punishment and his Judgement; disunity, rebellion, war and bloodshed, inflation, pestilence and many illnesses are prevailing (ibid., fols k5v–k6v).
77. Ibid., fol. k7r, fol. l1r.
78. Brazdil et al., 'Flood Events of Selected European Rivers in the Sixteenth Century', p. 255.
79. Christman, 'Heretics', p. 509.
80. C. Irenaeus, *Eine Christliche vnd tröstliche Predigt von Martha vnd Maria: Luce 10* (n.p. 1564), fol. A3r.

5 Irenaeus against the Concord: Polemical Use of *Wunderzeichen*, II

1. The difference between the view of Strigel and that of Flacius is carefully explained in L. Ilic, 'The Understanding of Sin in the Theology of Matthias Flacius', *Perichoresis*, 4:2 (2006), pp. 211–44; Christman, *Doctrinal Controversy*, pp. 49–60; and Christman, 'Heretics', pp. 71–87.
2. Christman, 'Heretics', p. 84.
3. Ilic, 'The Understanding of Sin', p. 219.
4. Christman, *Doctrinal Controversy*, p. 69.
5. Christman, 'Heretics', p. 99.
6. Christman, *Doctrinal Controversy*, p. 59.
7. The development of the Flacian Controversy in the Mansfeld region and the theological points of their conflicts are discussed in ibid., pp. 60–96.
8. Irenaeus's speech, which he delivered at the Eisleben Colloquium, is summarized in C. Spangenberg, *Historia: Warhafftige erzelung aller Geschichten wie wenn vnd warüber sich die Trennung vnter dern Predigern in der Graueschafft Mansfelt zugetragen* (Mansfeld, 1573), fols Nn3v–Nn4v.
9. Eisleben was now the headquarter of the anti-Flacius faction, but Wilhelm Sarcerius, who had become the pastor of the church in 1568, was sympathetic to the Mansfelder side and allowed Irenaeus to deliver a sermon. Christman, 'Heretics', p. 509.
10. C. Irenaeus, *Eine Hochzeit Predigt Vom Breutgam Christo vnd seiner Braut der Christlichen Kirchen* (Mansfeld, 1573).
11. C. Irenaeus, *Apostasia* (n.p., 1573), fols A4v–B1r.
12. Ibid., fol. S2v.
13. R. Kolb, *Confessing the Faith: Reformers Define the Church, 1530–1580* (St Louis, MO: Concordia, 1991), pp. 122–3.
14. *Exul Christi* was not a vain title for Irenaeus. In Mansfeld he found several comrades who had taken the same road of suffering. According to Christman, besides Irenaeus, four of the ministers in Mansfeld had experienced the loss of their posts in Saxony due to their opposition to the Synergist theology (Christman, 'Heretics', p. 231). Thus in Mansfeld Irenaeus met (or reunited with) a number of 'remnant' comrades who were willing to sacrifice their clerical careers for their faith. It is not difficult to imagine that this strengthened his determination to defend the Flacian position.
15. The full title in German is *Prognosticon Aus Gottes Wort notige Erinnerung Vnd Christliche Busspredigt zu diser letzten bosen Zeit An hohe vnd nider Standes Deutches Landes: Auff den Cometen so von Martini des 1577. Jars biss zurm Eyngang des 1578. Jars gesehen. Sampt Erzehlung vieler Cometen vnd anderer schrecklicher Zeichen vnd was allewegen darauff erfolget* (n.p., 1578).
16. The use of the title of *Prognosticon* for a religious message is not original to Irenaeus. Some authors exploited the popularity of astrological tracts and used the format of *Prognosticon* for their works, including those for religious instruction. For example, Urban Lüginslandt's *Spiritual Practica of the Year 1576 and until the End of this World* covers various daily concerns such as war and sickness and instructs readers how to deal with these ordeals; advice is not based on astrological analysis but on Scripture verses. U. Lüginslandt, *Geistliche Practica Von dem 1576. jaren bis zu end der Welt werende* (Basel, 1576).
17. Irenaeus, *Prognosticon*, fol. A3v.
18. Ibid., fol. B1r, fol. B2r, fol. B2v.
19. Ibid., fols D1r–D1v, fol. E1r, fols F1r–F1v, fol. F2r.

20. Ibid., fol. F4v. It seems that Irenaeus was not sure whether it was a comet or a new star.
21. Ibid., fols F4v–G1r.
22. Ibid., fols J3r–J4r.
23. For example, many phenomena were observed in Italy and other lands before the rise of the papacy around AD 600 (ibid., fol. K1v). In AD 612 a dreadful comet appeared, and a naked fiery sword stood in the sky. The sky in the north became red as if it was burning. Irenaeus states that these phenomena are signs of the rise of 'the devil's apostle of Mahomet' (ibid., fols K2v–K3r)
24. Ibid., fol. L1r.
25. Fincel, *Wunderzeichen*, fol. D7r.
26. Irenaeus, *Prognosticon*, fol. L1r.
27. Ibid., fol. M4r.
28. Ibid.
29. Fincel, *Wunderzeichen*, fols P7r–P7v.
30. Irenaeus, *Prognosticon*, fol. N2r.
31. Ibid., fol. N2v.
32. Ibid., fols O2v–O3r.
33. Ibid., fols O3r–O3v.
34. Ibid., fol. O3v. Probably he obtained this information from Spangenberg's Mansfeld chronicle.
35. Ibid., fol. R3r, fol. S2r, fol. S2v, fols S4r–S4v, fol. T3v.
36. Ibid., fol. V2v.
37. The Formula avoids using the term 'accident' and explains its doctrine of original sin without using this term. The Formula advises pastors not to use the words *substantia* and *accidens* in their sermons because they 'are not biblical terms and are words unfamiliar to common people'. R. Kolb and T. J. Wengert (eds), *The Book of Concord: The Confessions of the Evangelical Lutheran Church* (Minneapolis, MN: Fortress Press, 2000), p. 491.
38. Irenaeus, *Prognosticon*, fol. X2v.
39. Ibid., fol. X3r.
40. This is suggested by David's words: 'my sin is ever before me' and 'in sin my mother conceived me'; Psalm 51.
41. Irenaeus, *Prognosticon*, fols X3r–X3v.
42. Ibid., fols Y3v–Y4v.
43. Ibid., fol. Z1r.
44. C. Irenaeus, *Erklerung des Euangelij Luce XXI Am andern Sontag des Aduents* (n.p., 1581), fol. A4v, fol. C4r, fol. C4r.
45. Ibid., fol. C4v.
46. Irenaeus mentions several stories of monstrous births in the *Prognosticon* but did not add his personal comment on them.
47. C. Irenaeus, *Contrafet Vnd Spiegel des Menschen* (Ursel, 1582), fols B2v–B3r.
48. Ibid., fol. T5.
49. C. Irenaeus, *De Monstris: Von seltzamen Wundergeburten* (Ursel, 1584).
50. Ibid., fol. a4r, fol. a4v, fol. b2v.
51. Ibid., fol. b2v, fols b4v–c2v, fols e4a–e4b.
52. Ibid., fol. e2v, fol. f4r, fol. f4v.
53. Ibid., fols g4v–h1r, fol. h1r.
54. Ibid., fols J2v–J3r.

55. Ibid., fol. K2v. A photograph of this statue is printed in F. Martin, *Hellbrunn Castle near Salzburg* (Salzburg: Salzburger Verlag für Wirtschaft und Kultur, 1957), p. 26.
56. Irenaeus, *Monstris*, fol. S3v.
57. Ibid., fols 61r–61v, fol. 62r.
58. Ibid., fol. Cc4v, fol. Ee2v.
59. Ibid., fol. Cc4r.
60. Ibid., fol. Ff2r, fols Ff2r–Ff2v, fol. Gg1v, fol. Gg2r.
61. Ibid., fol. Hh2v.
62. Ibid., fols Hh3r–Hh3v.
63. Ibid., fols Pp4r–Qq1r. Christman (in *Doctrinal Controversy*) also points out that Mencel's position at the Weimar Colloquium in 1571 was close to the Flacian understanding.
64. Irenaeus, *Monstris*, fol. Ji4r.
65. Ibid., fol. Ll1r.
66. Ibid., fol. Jii2r.
67. Ibid., fols Jii2r–Jii2v.
68. Ibid., fol. Jii3r.
69. Ibid., fol. Kkk2r. After these arguments, at the end of the book, Irenaeus adds a new section and recounts two more stories of wonder. One is the news that a married couple in Bünigheim had fifty-three children, directly copied from a broadsheet published in Neuberg. The other is a story about a woman of Arnstadt who in 1569 bore two deformed babies who had claws like the devil's on their hands and feet. These stories have nothing to do with Irenaeus's discussion about conflicts between Christ and Belial, but probably he wanted to add the stories he had just obtained.
70. Brazdil et al., 'Flood Events of Selected European Rivers in the Sixteenth Century', pp. 260–1.
71. C. Koslofsky, *The Reformation of the Dead: Death and Ritual in Early Modern Germany, 1450–1700* (New York: St Martin's Press, 2000), pp. 128–9.
72. J. Aldenberger, *Fewer Wasser Wein Spiegel* (Leipzig, 1615), fol. a2v.
73. Ibid., fols Ff5v–Ff7v.
74. G. Rost, *Zweyfacher Theologischer Spiegel* (Rostock, 1625), fol. K3r.
75. Ibid., fol. M1v.
76. In the section of the year 1554, he recounts: 'people saw many *Wunderzeichen* as Caspar Goltwurm and others write. And in addition, (as *Jrenaeus* reports) sometime [in this year] a fiery red comet headed toward west from the sun. On February 19, two blood-red crosses stood in the sky'. Caesius, *Chronick*, fol. P7r. These cases are mentioned in Irenaeus, *Prognosticon*, fol. F3v, fol. N1r.
77. A. Bruckhart, *Cometenpredigt* (Madeburg, 1618), fol. D1v.

6 Andreae's Pastoral Use of *Wunderzeichen*

1. A. C. Piepkorn, 'Martin Chemnitz's Views on Trent: The Genesis and the Genius of the *Examen Concilii Tridentini*', *Concordia Theological Monthly*, 37 (1966), pp. 5–37, on p. 9.
2. R. Kolb, 'Jakob Andrae's Concern for the Laity', *Concordia Journal*, 4 (1978), pp. 58–67, on pp. 58 and 61.
3. M. J. Haemig, 'The Living Voice of the Catechism: German Catechetical Preaching 1530–1580' (PhD dissertation, Harvard Divinity School, 1996); M. J. Haemig, 'Laypeople as Overseers of the Faith: A Reformation Proposal', *Trinity Seminary Review*, 27:1 (2006), pp. 21–7.

4. See Rittgers, *The Reformation of Suffering*.
5. Spangenberg's pastoral efforts for the Mansfeld miners is well discussed in S. Boettcher, 'The Rhetoric of "Seelsorge" for Miners in the Sermons of Cyriakus Spangenberg', in G. Litz, H. Munzert and R. Liebenberg (eds), *Frömmigkeit- Theologie-Frömmigkeitstheologie: Contributions to European CH Festschrift für Berndt Hamm* (Leiden: Brill, 2005), pp. 453–66.
6. J. Andreae, *Vier Christliche Predigten Vber etliche Euangelia im Aduent* (Regensburg, 1588), fol. S3v, fols S3v–S4r, fol. S4r.
7. Though I could not consult the original, according to Charlotte Methuen, Andreae mentions celestial wonder in his sermon of 1581 and states that the nova in 1572 and the comets of 1577–8 and 1581 signify God's wrath urging people to amend their lives. Though the list of signs became longer, again Andreae mentions only phenomena witnessed by everyone. Methuen points to Andreae's negative assessment of astrology and admits that he 'viewed the study of nature more with suspicion than with enthusiasm'; Methuen, *Kepler's Tübingen*, pp. 128–9, p. 220.
8. Andreae, *Vier Christliche Predigten*, fols S4v–T3v.
9. J. Andreae, *Christliche notwendige vnd ernstliche Erinnerung Nach dem Lauff der jrrdischen Planeten gestalt: Darauss ein jeder einfältiger Christ zusehen was für Glück oder Vnglück Teustchlandt diser zeit zugewarten* (Tübingen, 1568).
10. Andreae called his book a New Year's gift; ibid., fol. Cc1r.
11. Ibid., fol. Aa2v.
12. P. Severus, *Newe Zeitung: Von den Bedeutung die da folgen werden aus den Finsternissen Welche zum teil gesehen sind vnd noch zukünfftig geschehen sollen* (n.p., 1567), fols A2v–A4r.
13. Ibid., fol. A4r.
14. A. Scultetus, *Prognosticon Nouilunij Ecliptici* (Görlitz, 1567), fol. F4r, fol. J1v.
15. Andreae, *Erinnerung*, fol. Aa3r, fols Aa3r–Aa4r, fol. Bb2v.
16. Ibid., fols Bb3r–Bb3v, fol. Cc2r, fols Cc3r–Cc3v, fols Cc3v–Cc4v, fol. Dd3v, fols Ee1r–Ee1v.
17. Ibid., fol. A2r, fols A2r–A2v, fols A3r–A3v.
18. Ibid., fols A4v–B1r, fol. J2r, fols G1v–G2r, fol. J2r, fols S2v–S3r, fols S4r–S4v.
19. Ibid., fol. J2v.
20. Methuen, *Kepler's Tübingen*, p. 45.
21. J. Andreae, *Ein Christliche Predig Vber das Euangelium auff den xxiiij Sontag nach Trinitatis Matthei am 24 Von vieler vnd mancherley verfürungen in der Kirchen Gottes vor dem Jüngste tage* (Leipzig, 1578), fols B2r–B2v.
22. Ibid., fols D1r–D2v, fols D2v–D4r, fols D4v–E1r, fol. E1r, fols E2r–E3r.
23. Kolb, 'Andreae's Concern', p. 65.
24. Andreae, *Christliche Predig*, fol. F2v.
25. Ibid., fols J3v–J4r, fol. K2v, fols K3r–K4r.
26. Ibid., fols K4r–K4v, fols L1r–L2r.
27. Ibid., fol. K2v.

7 Celestial Wonders under the Shadow of War

1. J. Sifard, *Cometen Spiegel* (Leipzig, 1605), fol. A2v.
2. Cited in Edward H. Thompson's translation of J. V. Andreae, *Christianopolis* (Dordrecht: Kluwer, 1999), p. 251.
3. B. Nischan, 'The Palatine and Brandenburg's "Second Reformation"', in D. Visser (ed.), *Controversy and Conciliation: The Reformation and the Palatine 1559–1583* (Allison Park, PA: Pickwick Publications, 1986), pp. 303–18.
4. G. Parker, *The Thirty Years' War* (New York: Military Heritage Press, 1987), p. 6.
5. Ibid., p. 12.
6. Ibid., pp. 12–13.
7. H. Leuchter, *Ein Christlich Predigt: Von dem erschrecklichen Chasmate, welchs am 7. Nouemb. dieses 1605. Jahrs oben am Himmel ... gesehen worden* (Darmstadt, 1605), fol. A2v. Leuchter also published a short book on wondrous signs in 1613; see H. Leuchter, *Discurß Von etlichen Zeichen* (Darmstadt, 1613), fol. A3r, fol. B4r, fol. B4v.
8. S. Hering, *Eine guthertzige warnungsschrifft für künfftigem Vngluck vnsers lieben Vaterlandes Deutscher Nation* (Dresden, 1609), fol. N1v, fols N1v–N2r.
9. Ibid., fol. N1v.
10. Parker, *The Thirty Years' War*, pp. 8–9.
11. H. Braun, *Der heßlich vnd greßliche fewrige Drache* (Darmstadt, 1616), fols A2v–A3v, fols C4v–D1r, fol. A3v, fol. A4v.
12. Ibid., fol. D1r.
13. Braun also published a tract on the three suns observed in March 1615, but the contents follow the conventional pattern of *Wunderzeichen* discourses: H. Braun, *Die drey Sonnen* (Wittenberg, 1615).
14. For example, in his *Postilla*, Conrad Schlüsselburg refers to Copernicus's *libri reuolutionum* and states that the sun had come 2,666 miles nearer to the earth since the time of Ptolemy. Schlüsselburg, Superintendent of Stralsund, believed that this could be included among the signs of the world's end mentioned by Christ. Schlüsselburg, *Postilla*, pp. 46–7.
15. The impact of Arndt's book on Lutherans of his age and later generations is discussed in J. Wallmann, 'Johann Arndt (1555–1621)', in C. Lindberg (ed.), *The Pietist Theologians: An Introduction to Theology in the Seventeenth and Eighteenth Centuries* (Malden: Blackwell, 2005), pp. 21–37, on pp. 21–2, pp. 33–5.
16. J. Arndt, *True Christianity*, trans. P. Erb (New York: Paulist Press, 1979), p. 21.
17. Ibid., p. 174.
18. J. Arndt, *Of True Christianity Four Books* (London, 1714), pp. 464–5, pp. 477–8, p. 478, p. 480, p. 480.
19. J. Arndt, *Postilla* (Jena, 1616), fol. D1r, fol. D1v, fol. D2r.
20. Barnes, *Prophecy and Gnosis*, p. 257.
21. Arndt, *Postilla*, fol. C3r.
22. F. A. Yates, *The Rosicrucian Enlightenment* (London: Routledge, 1971), p. 256.
23. Ibid., p. 256. A. Åkerman examines the relations between astronomical phenomena and the Rosicrucian manifestos in 'The Rosicrucians and the Great Conjunctions', in J. C. Laursen and R. H. Popkin (eds), *Continental Millenarians: Protestants, Catholics, Heretics* (Dordrecht: Kluwer, 2001), pp. 1–8.
24. Yates, *The Rosicrucian Enlightenment*, p. 255.
25. This illustration is printed in the frontispiece of ibid.

26. Åkerman, 'The Rosicrucians and the Great Conjunctions', pp. 4–7; Yates, *The Rosicrucian Enlightenment*, pp. 34–5; Barnes, *Prophecy and Gnosis*, pp. 224–5; D. R. Dickson, 'Johann Valentin Andreae's Utopian Brotherhoods', *Renaissance Quarterly*, 49:4 (1996), pp. 760–98, on p. 769.
27. Dickson analyses the relationship between these Rosicrucian writings, J. V. Andreae and his dream of Protestant brotherhood in 'Johann Valentin Andreae's Utopian Brotherhoods'.
28. Andreae, *Christianopolis*, p. 239.
29. On the messages of *Wetterpredigten*, see my forthcoming article: K. Kurihara, '"Whether We Live or Die, We Are the Lord's": Lutheran *Wetter* Discourses and Spiritual Interpretations of Natural Disasters', in K. A. Edwards (ed.), *The Science of the Supernatural in Early Modern Europe* (New York: Palgrave Macmillan, forthcoming).
30. D. Bramer, *Vom Donner Blitz Hagel Sturmwinden vnd andern grossen Vngewittern* (Erfurt, 1577), fols E7v–F5r.
31. For example, the Horneburg minister Johann Richter's sermon on storms (1624); see J. Risler, *Himmels Predigten Von Donner vnnd Blitz* (Hamburg, 1624), p. 63.
32. Soergel, *Miracles*, p. 174.
33. Sifard, *Cometen Spiegel*, p. 72.
34. J. Mülmann, *Threni Iesv Christi, Das ist Threnenpredigten … Mit Angehenckter Bußpredigt von dem schrecklichen Fewerzeichen* (Leipzig, 1605), pp. 61–2.
35. Coler, *Eigentlicher bericht*, fol. C1v. Luther's open attitude towards the angels' possible roles as guardians of believers had to encourage Lutherans' belief in angelic encounters. See P. M. Soergel, 'Luther on the Angels', in P. Marshall and A. Walsham (eds), *Angels in the Early Modern World* (Cambridge: Cambridge University Press, 2006), pp. 64–82.
36. I would like to express my deep gratitude to Jürgen Beyer for sharing with me his manuscript on the Lutheran lay prophets, which contains the information about Anna's case quoted here. The English translation of the description of the crowd's reactions is Beyer's.
37. D. Meder, *Warhafftige beschreibung eines grausamesn erchröcklichen Grossen Wasserflut* (Nuremberg, 1589), fol. B1v.
38. Schütz, *Newe Zeytung*.
39. Anon., *Ein schönes neues Klaglied von der letzten Zeit* (Laingen, 1619).
40. J. Theibault, 'Jeremiah in the Village: Prophecy, Preaching, Pamphlets, and Penance in the Thirty Years' War', *Central European History*, 27:4 (1994), pp. 441–60, on p. 441.
41. C. Dietrich, *Ulmische Cometen Predigte* (Ulm, 1619), fols B1r–B1v, fol. D2v.
42. Ibid., fol. F2v.
43. T. Kaufmann, 'The Lutheran Sermon during the War and at the Time of the Peace Agreement', in K. Bussmann and H. Schilling (eds), *1648: War and Peace in Europe*, 3 vols (Münster: Westfälisches Landesmuseum, 1998), vol. 1, pp. 245–50, on p. 248.
44. Quoted in B. von Krusenstjern, 'Prodigienglaube und Dreißigjähriger Krieg', in H. Lehmann and A.-C. Trepp (eds), *Im Zeichen der Krise: Religiosität im Europa des 17. Jahrhunderts* (Göttingen: Vanderhoeck & Ruprecht, 1999), pp. 53–78, on p. 75.
45. Ibid., p. 74.
46. H. Berg, *Military Occupation under the Eyes of the Lord: Studies in Erfurt during the Thirty Years War* (Göttingen: Vandenhoeck & Ruprecht, 2010), pp. 109–10.
47. Ibid., p. 110. Fritz started writing his chronicle *Cosmographia* between 1665 and 1676.
48. Theibault, 'Jeremiah in the Village', pp. 453–4. Also see G. Mortimer, *Eyewitness Accounts of the Thirty Years War 1618–48* (New York: Palgrave, 2002), pp. 180–1.
49. *Ein warhafftige Newe Zeitung. Vnd erbärmliches Wunderwerck: wie vns Gott ein schöne Ruthen an den Himmel gesiecht solliche ist gesehen worden* (Augsburg, 1621).

50. Krusenstjern, 'Prodigienglaube und Dreißigjähriger Krieg', p. 77.
51. Ibid., p. 78.
52. Mortimer, *Eyewitness Accounts*, p. 19.
53. Berg, *Military Occupation*, pp. 9–10.
54. Barnes, *Prophecy and Gnosis*, pp. 239–40.
55. Wilhelm Schmidt-Biggemann summarizes the prophecies of Böhme and some others in 'The Apocalypse and Millenarianism in the Thirty Years' War', in K. Bussmann and H. Schilling (eds), *1648: War and Peace in Europe*, 3 vols (Münster: Westfälisches Landesmuseum, 1998), vol. 1, pp. 259–63.
56. Pierre Bayle's book on comets, inspired by the 1680 comet, appeared in 1682. P. Bayle, *Various Thoughts on the Occasion of a Comet* (1682), trans. R. C. Bartlett (Albany, NY: State University of New York Press, 2000).

WORKS CITED

Primary Sources

Aldenberger, J., *Fewer Wasser Wein Spiegel* (Leipzig, 1615).

Alexander, D. (ed.), *The German Single-Leaf Woodcut, 1600–1700*, 2 vols (New York: Abaris Books, 1977).

Amsdorf, N. von, *Fünff furnemliche vnd gewisse Zeichen aus heiliger göttlicher Schrifft* (Jena, 1554).

Andreae, J., *Christliche notwendige vnd ernstliche Erinnerung Nach dem Lauff der jrrdischen Planeten gestalt* (Tübingen, 1568).

—, *Ein Christliche Predig Vber das Euangelium auff den xxiiij Sontag nach Trinitatis* (Leipzig, 1578).

—, *Vier Christliche Predigten Vber etliche Euangelia im Aduent* (Regensburg, 1588).

—, *Andreae and the Formula of Concord: Six Sermons of the Way to Lutheran Unity*, ed. and trans. R. Kolb (St Louis, MO: Concordia, 1977).

Andreae, J. V., *Christianopolis*, trans. E. H. Thompson (Dordrecht: Kluwer, 1999).

Anon., *Etliche gesichte sozu Holtzhausen vnther Wasserburg im Lande During gelegen am Donnerstag nach Trinitatis Und zu Pretin den xx. Junij Diese xlviij Jars am Himmel von glaubwirdigen Leuten seint gesehen worden* (n.p., 1548).

—, *Ein Gesicht am Himel gesehen: Wie das Götliche Gericht Am tag Visitations Anno 1556* (n.p., 1556).

—, *Sehr grewliche erschrockliche vor vnerhörte warhafftige Newe Zeytung: was für grausame Tyranney der Moscowiter an den Gefangen hinweggefürten Christen auss Lyffland* (Nuremberg, 1561).

—, *Sechs schöne newe Geistliche Lieder* (Lübeck, 1580).

—, *Warhafftige newe Zeitung vnd erschreckliche Wunderzeichen so sich newlichen zu Rom vnnd Paris in dem nechst vergangen 83. Jar geschehen* (Augsburg, 1584).

—, *Zwo Warhafftige Newe Zeittung Vnd gründliche Geschicht* (Graz, 1593).

—, *Ein schönes neues Klaglied von der letzten Zeit* (Laingen, 1619).

Arndt, J., *Postilla* (Jena, 1616).

—, *Of True Christianity Four Books* (London, 1714).

—, *True Christianity*, trans. P. Erb (New York: Paulist Press, 1979).

Bapst, M., *Warhafftige beschreibung Des Erschrecklichen Blut vnd Fewerzeichens welches den 6. Martij dieses ablauffenden 1582. Jahrs am Himmel gesehen worden* (Freiberg, 1583).

Baumgarten, J., *Beschreibung Der Schrecklichen Brunst Fewersklufft Chasmatis vnd Gesichtes so den 27 Augusti des jetzt lauffenden 1581. Jars* (Magdeburg, 1581).

Bayle, P., *Various Thoughts on the Occasion of a Comet* (1682), trans. R. C. Bartlett (Albany, NY: State University of New York Press, 2000).

Bramer, D., *Vom Donner Blitz Hagel Sturmwinden vnd andern grossen Vngewittern* (Erfurt, 1577).

Braun, H., *Die drey Sonnen* (Wittenberg, 1615).

—, *Der heßlich vnd greßlich fewrige Drache* (Darmstadt, 1616).

Bruckhart, A., *Cometenpredigt* (Madeburg, 1618).

Caesius, G., *Chronick Oder ordenliche verzeichnuss vnnd beschreibung aller Cometen* (Nuremberg, 1579).

Carion, J., *Außlegung der verborgenen Weissagung D. Johan. Carionis von veranderung und zuvfelligem glück der hösten Potetaten des Römischen Reichs* (Strasbourg, 1548).

Celichius, A., *Christliche Notwendige nutzliche vnd Theologische erinnerung von dem newen Cometen* (Leipzig, 1578).

—, *Theologische erinnerung von dem newen Cometen* (Magedeburg, 1578).

Chemnitz, M., *Postilla* (Magdeburg, 1594).

—, 'An Autobiography of Martin Kemnitz', trans. A. L. Graebner, *Theological Quarterly*, 3 (1899), pp. 472–87.

—, N. Selnecker and T. Kirchner, *Refutatio Irenaej: Grundlicher Bericht auff das Examen M. Christophori Irenej, so er Anno 1581. wider den ersten Artickel deß Christilichen ConcordiBuchs* (Heidelberg, 1583).

Chytraeus, D., *Vom Newen Stern Welcher Anno M.D.LXXII. Im Nouember erschienen. Vnd vom Cometen Welcher wir im Nouember dieses lauffenden M.D.LXXVII. Jars vnd noch jtzund sehen: Erinnerung* (Rostock, 1577).

Coler, J., *Notwendige Erinnerung auff dz schreckliche Fewerzeichen So den 10. Septembris dess jetzlauffenden 1580 Jahrs am Himmel gesehen worden* (Berlin, 1581).

—, *Eigentlicher bericht vo[n] den setzamen vnd zu vnser Zeit vnerhörten Wunderwercken vnd Geschichten so sich newlicher zeit in der Marck Brandenburg zugetragen vnd verlauffen haben vnd noch teglich geschehen* (Erfurt, 1595).

Creat, J., *Kurtze beschreibunge des Cometen welcher ist gesehen worden am Himmel Anno 1577. den 11. Nouembris Auch von etlichen Wunderzeichen die vorher gegangen sein* (n.p., [*c.* 1577]).

Cuno, J., *Erinnerung Von dem schrecklichen Fewerschuss Welcher am andern Tage des Herbstmonats früe Morgens zwischen 2. Vnd 3. Vhr gesehen worden* (Eisleben, 1579).

Dietrich, C., *Ulmische Cometen Predigte* (Ulm, 1619).

Eber, P., *Beschreibung des schrecklichen zeichens so am 13. tag Martij fast die gantze nacht vber zu Wittenberg vnd an viel andern orten ist gesehen worden mit einer vermanung* (Wittenberg, 1562).

Erasmus, D., *Ten Colloquies of Erasmus*, ed. C. R. Thompson (New York: Liberal Arts Press, 1957).

Erben, N., *Zur Buss vermanende Comet: Anno M.D.LXXII. Den XIIII. Maij den abendt vmb VIII. Vhr erschienen* (Erfurt, 1582).

Faber, B., *Allerley Christliche nötige vnd nützliche vnterrichtungen von den Ketzen Hendeln der Welt* (Eisleben, 1565).

Fagius, C., *Wunder Stern vnd Zornzeichen: So an Sonn vnd Monde des 1568. Jars den. 14. 21. vnd 22. Tag Decembris zu Erffordt vnd angrentzenden örtern gesehen worden* (Erfurt, 1568).

Fincel, H., *Wunderzeichen* (Jena, 1556).

—, *Der dritte theil Wunderzeichen* (Frankfurt am Main, 1567).

Flacius Illyricus, M., *Eine Weissagung vnd ein schöner Herrlicher trost für alle hochbetrübte frome Christliche hertzen zu diser jtzigen trübseligen zeit: Aus dem XIII. Cap. Der offenbarung Johannis* (Magdeburg, 1548).

—, *Kleriche beweisung das alle die jenige welche die schrifften widder das INTERIM vnd Mitteldingfeil zuhaben vn[d] zu lesen verbieren* (Magdeburg, 1550).

—, *Ordenung vnd Mandat Keiser Caroli V. vernewert im April Anno 1550* (Magdeburg, 1550).

—, *Gründliche Verlegung des Langen Comments der Adiaphoristen oder der verzelung jrer handlungen: Zu gründlicher erforschung der warheit in dieser sache sehr nützlich zu lessen* (Jena, 1560).

Gedike, S., *Postilla* (Eisleben, 1588).

Gernhard, B., *Vom Jüngsten Tage: Vier nützliche Predigten inn Zwölff Heuptartickel* (Erfurt, 1556).

Gölitz, J., *Newe Zeitung: Ein Wunderbarlich Gesicht am Himmel so durch den Pfarrhen vnnd ander Personen des Dorffs Aschre inn der Löblichen Herrschafft Thoma gelegen gesehen ist worden* (n.p., 1560).

Goltwurm, C., *Wunderwerck vnd Wunderzeichen Buch* (Frankfurt am Main, 1557).

Harms, W. (gen. ed.), *Deutsche Illustrierte Flugblatter des 16. und 17. Jahrhunderts* (Tübingen and Munich, 1980–).

—, M. Schilling, B. Bauer and C. Kemp (eds), *Deutsche Illustrierte Flugblatter des 16. und 17. Jahrhunderts, Volume 1: Die Sammlung der Herzog August Bibliothek in Wolfenbüttel, Teil 1. Ethica and Physica* (Tübingen: Max Niemeyer, 1985).

—, and C. Kemp (eds), *Deutsche Illustrierte Flugblatter des 16. und 17. Jahrhunderts, Volume 4: Die Sammlungen der Hessischen Landes- und Hochschulbibliothek in Darmstadt* (Tübingen: Max Niemeyer, 1987).

—, and M. Schilling (eds), *Deutsche Illustrierte Flugblatter des 16. und 17. Jahrhunderts, Volume 7: Die Sammlung der Zentralbibliothek Zürich Teil 2: Die Wickiana II (1570–1588)* (Tübingen: Max Niemeyer, 1997).

—, and — (eds), *Deutsche Illustrierte Flugblatter des 16. und 17. Jahrhunderts, Volume 6: Die Sammlung der Zentralbibliothek Zürich Teil 1: Die Wickiana I (1500–1569)* (Tübingen: Max Niemeyer, 2005).

Hebenstreit, J., *Wunderzeichen* (Erfurt, 1564).

—, *Prognosticon Historicvum Auffs 1568 Jhars* (Erfurt, 1567).

Heerbrand, J., *Ein Predig Von dem erschroklichen Wunderzeichen am Himmel dem newen Cometen oder Pfawenschwantz Gehalten zu Tübingen den 24. Sontag nach Trinitatis wolcher ist der 17. Wintermonatas* (Tübingen, 1577).

Heisterbach, C. von, *The Dialogue on Miracles* (London: Routledge, 1929).

Hering, S., *Eine guthertzige warnungsschrifft für künfftigem Vngluck vnsers lieben Vaterlandes Deutscher Nation* (Dresden, 1609).

Hirschmann, G. (ed.), *Die Kirchenvisitation in Landgebiet der Reichsstadt Nürnberg 1560 und 1561: Quellenadition* (Neustadt an der Aisch: Degener, 1994).

Hunnius, A., *Postilla* (Wittenberg, 1592).

Irenaeus, C., *Symbolvm Apostolicvm: Das ist Die Artickel vnsers Christlichen Glaubens ausgelegt. Mit Vermeldung vnd widerlegung allerley Jrthumb vnd Ketzerien beide alt vnd new so darwider entstanden* (Eisleben, 1562–3).

—, *Abdruck: Eines schrecklichen Zornzeichens Sampt Christlicher vnd notiger Erinnerung* (Eisleben, 1564).

—, *Eine Christliche vnd tröstliche Predigt von Martha vnd Maria: Luce 10* (n.p., 1564).

—, *Wasser-Spiegel* (Eisleben, 1566).

—, *Warnung vnd vrsachen: Das man nicht in eine Amnistiam vnd Stilschweigen der Jrthumen vnd Corruptelm Noch in den Orden der newen Jacobsbrüder willigen soll* (n.p., 1569).

—, *Apostasia* (n.p., 1573).

—, *Eine Hochzeit Predigt Vom Breutgam Christo vnd seiner Braut der Christlicher Kirchen* (Mansfeld, 1573).

—, *Prognosticon Aus Gottes Wort nötige Erinnerung Vnd Christliche Busspredigt zu diser letzeten bosen Zeit* (n.p., 1578).

—, *Erklerung des Euangelij Luce XXI Am andern Sontag des Aduents* (n.p., 1581).

—, *Contrafet Vnd Spiegel des Menschen* (Ursel, 1582).

—, *De Monstris: Von seltzamen Wundergeburten* (Ursel, 1584).

—, *Euangelischer Vnd tröstlicher Gnadenspiegel Wieder den Schrecklichern Zornspiegel des Gesetzes* (Ursel, 1593).

Kolb, R., and T. J. Wengert (eds), *The Book of Concord: The Confessions of the Evangelical Lutheran Church* (Minneapolis, MN: Fortress Press, 2000).

Krentzheim, L., *Coniecturae: Christliche vermuttungen von künfftiger Zeit Zustandt in Kirchen vnd Regimenten* (Görlitz, 1583).

Leuchter, H., *Ein Christlich Predigt: Von dem erschrecklichen Chasmate, welchs am 7. Nouemb. dieses 1605. Jahrs oben am Himmel ... gesehen worden* (Darmstadt, 1605).

—, *Discurß Von etlichen Zeichen* (Darmstadt, 1613).

Licht, G., *Von dem schröcklichen vnd betrübtem Nawen* [*sic*] *Jhar so an den Wolcken des Himels den. 38.* [*sic*] *Decembris dieses 61. Jhars gesehen* (Frankfurt an der Oder, 1561).

Lügislandt, U., *Geistliche Practica Von dem 1576. jaren bis zu end der Welt werende* (Basel, 1576).

Luther, M., *Dr. Martin Luthers Sämmtliche Schriften*, ed. J. G. Walch, 25 vols (St Louis, MO: Concordia, 1887).

—, *Luther's Works*, 55 vols (St Louis, MO: Concordia, 1957–86).

—, *Sermons of Martin Luther*, ed. and trans. J. N. Lenker, 8 vols (Grand Rapids, MI: Baker Books, 1983).

—, *Sermons of Martin Luther: The House Postils*, ed. and trans. E. F. A. Klug, 3 vols (Grand Rapids, MI: Baker Books, 1996).

Lyscothenes, C., *Wunderwerck Oder Gottes vnergründliches vorbilden* (Basel, 1557).

Meder, D., *Zehen Christliche Busspredigten vber Die Weissagung Christi dess grossen Propheten vom Ende der Welt vnd Jüngste Tage* (Frankfurt am Main, 1581).

—, *Warhafftige beschreibung eines grausamesn erchröcklichen Grossen Wasserflut* (Nuremberg, 1589).

Medler, N., and M. Flacius Illyricus, *Zeychen am himmel bey Braunschwig newlich gesehen: durch den superatendentem zu Braunschwig geschriben Mit einer vorrede* (Magdeburg, 1549).

Melanchthon, P., *Corpus Reformatorum. Philippi Melanthonis opera quae supersunt Omnia*, ed. K. Gottlieb Bretschneider, 28 vols (Halle: C. A. Schwetschke, 1840).

—, *Orations on Philosophy and Education*, ed. S. Kusukawa (Cambridge: Cambridge University Press, 1999).

—, and M. Flacius Illyricus, *Ein grausam Meerwunder den Bapst bedeutende zu Rom gefunden* (Magdeburg, 1550).

Mencel, H., *Postilla* (Leipzig, 1596).

Mirus, M., *Postilla* (Jena, 1605).

Monner, B., *Bedencken vonn dem Kriege der Anno Sechs siben vnd viertzig im land zü Meissen vnd Sachsen gefürt ist so für er zühalten sey gestalt* (Basel, 1557).

Müller, S., *Astronomische Beschreibung dess Cometen so zu ende dises verloffenen 1577. Jars erscheinen sampt seiner bedeutung* (Augsburg, 1577).

Mülmann, J., *Threni Iesv Christi, Das ist Threnenpredigten ... Mit Angehenckter Bußpredigt von dem schrecklichen Fewerzeichen* (Leipzig, 1605).

Musculus, A., *Weissagung D. Martini Lutheri der Deutschen Apostel vnd Prophet von dem grossen zunahenden vnglück über Deutschland durch den Gog Magog vnd Mensch* (Frankfurt an der Oder, 1568).

—, *Gebet Vmb abwendung woluerdienter zunahende Straff vber Deutschland* (Erfurt, 1569).

—, *Vom Mensch vnd Kedar, vom Gog vnd Magog* (Frankfurt an der Oder, 1577).

Opitz, H., *Von dem erschrecklichen vnnd grossen Fewrigen Zeichen welches am Himel am tag der vnschüldigen Kindtlein im Jar nach der geburt Christi M.D.LXI. An vielen orten vnd Stedten ist gesehen worden* (Nuremberg, 1561).

Orphanus, N., *Beschreibung des Erschrecklichen Brennenden Flammenden vnd Stralschiessenden Fewers vnd Zornzeichen Gottes vber Europa* (Eisleben, 1575).

Pauli, S., *Postilla. Theil 1* (Magdeburg, 1572).

Peristerus, W., *Eine Trewhertzige Bußpredigt Vber den Straffspruch Christi aus dem Euangelio des 21. Sontags nach Trinitatis* (Berlin, 1580).

—, *Eine Trostreiche Predigt Vber das Euangelium Vom Cananeischen Weiblein ... Sampt einer kurtzen Erinnerung vnd Erklerung deß schrecklichen Wunderzeichen so den Sonabend zuuor nach Mittage erschienen vnnd gesehen ist worden* (Berlin, 1581).

Pfeffinger, J., *Christliche Gewisse Deutung der Zeichen die für vnd in diesem 1562. Jar gesehen* (Leipzig, 1562).

Pilich, J., *Eine Predigte Am newen Jarstage mit kürtzer vermanung zur Buße vnd bekerung* (Magdeburg, 1581).

Pletzlein, W., *Warhafftige Bericht Vnd Kurtze Beschreibung des Grossen vnnnd Erschrecklichen Wunderzeichens So sich am heiligen Ostertag den 19. Aprils dieses jtzlauffenden 84. Jars an der Sonnen begeben* (Hof, 1584).

Praetorius, A., *Schade Wacht: Nothwendige Warnung vnd Vermanung an alle busfertige glaubide Christen getan von dem erschrecklichen Comet vnd Zornzeichen der dieses 77. Jar am Himmel gesehen wird* (Erfurt, 1577).

—, *Selige Erinnerung Von dem erschreklichen Fewerenden Cometstern am Himel* (Erfurt, 1580).

Renner, J., *Johannes Renner's Livonian History 1556–1561*, trans. J. S. Smith and W. Urban (Lewiston, NY: Edwin Mellen Press, 1997).

Risler, J., *Himmels Predigten Von Donner vnnd Blitz* (Hamburg, 1624).

Rost, G., *Zweyfacher Theologischer Spiegel* (Rostock, 1625).

Saccus, S., *Erklerung Vber die Sontags Euangelia vnd vber die fürnembsten Fest durchs gantze Jahr* (Magdeburg, 1592).

Schaller, D., *Theologischer Heroldt* (Magdeburg, 1604).

Scheitlich, J., *Warnung für der grossen erschrecklichen vnd nun fast zu nahender straffen Gottes vber das gantze Deustchland* (Eisleben, [*c.* 1565]).

Schlüsselburg, C., *Postilla* (Frankfurt am Main, 1604).

Schomer, C., *Sieben Bußpredigten Von den zukünfftigern schrecklichen straffen großen verenderung vnd grewlichen verhanden Trübsalen so für dem Jüngsten tage fürher gehen vnd sonderlich vber Deutschland kommen werden* (Lemgo, 1583).

Schütz, J., *Newe Zeytung: Was sich allhie im Landt zu Sachssen hat zugetragen in der Stadt Bitterfeldt an an* [*sic*] *der Dam wie ein menschliche Handt mit einem Blut rhoten Schwerdy am Hellen Himel gestanden ist vnd Blut vom Himel gefallen ist. Anno 1568* (n.p., 1568).

—, *Von den Wunderzeichen am Himel vor dem grossen tages des Herr aus der Sprüh des Propheten Joel am andern Capittel* (Wittenberg, 1582).

Scultetus, B., *Prognosticon Nouilunij Ecliptici* (Görlitz, 1567).

Selnecker, N., *Ein Christlich Gebet in jetzigem eleneden zustand darin Gott selbs vns seine fewrige Rute vnd seinenn gerechten Zorn am Himmel zeiget vnd vns mit allerley drawungen vnd züchtigungen heimsuchet zuthun vnd sich damit für Gott demütiglich anzugehen vmb Gnad zu bitten* (Leipzig, 1578).

Severus, P., *Newe Zeitung: Von den Bedeutung die da folgen werden aus den Finsternissen Welche zum teil gesehen sind vnd noch zukünfftig geschehen sollen* (n.p., 1567).

Sifard, J., *Cometen Spiegel* (Leipzig, 1605).

Spangenberg, C., *Ernste vnd hochnöttige Busspredigt An gantzes Deutschland* (Eisleben, 1569).

—, *Mansfeldische Chronica: Der Erste Teil* (Eisleben, 1572).

—, *Historia: Warhafftige erzelung aller Geschichten wie wenn vnd warüber sich die Trennung vnter dern Predigern in der Graueschafft Mansfelt zugetragen* (Mansfeld, 1573).

Spangenberg, J., *Postilla Vom Aduent bis auff Ostern: Für die jungen Christien Knaben vnd Meidlein in Fragstücke verfasset* (Erfurt, 1564).

Stoltz, J., and J. Aurifaber, *Kurtze Verlegung der vnchristlichen Practica Magistri Johannis Hebestreits auff das jar 1554 zu Erffurd ausgangen* (Jena, 1554).

Strauss, W. L. (ed.), *The German Single-Leaf Woodcut, 1550–1600*, 3 vols (New York: Abaris Books, 1977).

Suevus, S., *Cometen: Was sie für grosse Wunder vnd schreckliche ding zu bedeuten vnd anzukündigen pflegen* (Görlitz, 1578).

—, *Erbauungsschriften: Spiegel des Menschlichen Lebens*, ed. M. A. van den Broek (Amsterdam: Rodopi, 1984).

Taurer, A., *Hochnotwendigster Bericht Von mancherley erschreckliche Wunderzeichen vnd Bußruffern* (Halle, 1592).

Ursinus, A., *Kurtze Beschreibunge der geschehen vnd gesehenen vnnatürlichen Wunderzeichen am Himmel im 1568. 69. vnd 70. Jhare* (Erfurt, 1570).

—, *Prognostication Auff das Jhar nach der Geburt Jhesu Christi vnsers Heylandes. M.D.LXXIIII: Beyneben einer kurtzer Beschreibung des erschienen Cometens im 1572. vnd 1573 Jhare* (Erfurt, 1574).

Victorius, P., *Xasmatodogia Oder Erinnerung von der schrecklicher Chasmate oder Fewerzeichen* (Magdeburg, 1583).

Vischer, C., *Außlegung der Euangelien: so man auff die Sontagen in der Christliche Kirchen zu handeln pfleget* (Schmalkalden, 1570).

Walther, G., *Außlegung der Euangelien so an den Sontagen gelesen werden. Theil. 1.* (Wittenberg, 1579).

Weller, H., *Wunderliche vnd vnerhorte Geschicht die sich zu Freybergl vnnd inn derselben gegent inn Meissen zugetragen hat den xiij. Augusti / Jm M.D.Lix* (Magdeburg, 1559).

Wigand, J., *Vom Straffampt der Sünden* (Ursel, 1564).

—, *Postilla* (Frankfurt am Main, 1569).

—, *Vom Flickwerck M. Jrenaej* (Jena, 1572).

Wittich, J., *Gründtliche vnd warhafftige Bericht. Von dem Erschrocklichen vnd Wunderbarlichen Zeichen* (Erfurt, 1561).

Secondary Sources

Åkerman, A., 'The Rosicrucians and the Great Conjunctions', in J. C. Laursen and R. H. Popkin (eds), *Continental Millenarians: Protestants, Catholics, Heretics* (Dordrecht: Kluwer, 2001), pp. 1–8.

Anderson, M. S., *The Origins of the Modern European State System, 1494–1618* (London: Longman, 1998).

Arand, C. P., J. A. Nestingen and R. Kolb, *The Lutheran Confession: History and Theology of the Book of Concord* (Minneapolis, MN: Fortress Press, 2012).

Barker, P., 'The Role of Religion in the Lutheran Response to Copernicus', in M. J. Osler (ed.), *Rethinking the Scientific Revolution* (Cambridge: Cambridge University Press, 2000), pp. 59–88.

Barnes, R. B., *Prophecy and Gnosis: Apocalypticism in the Wake of the Lutheran Reformation* (Stanford, CA: Stanford University Press, 1988).

Behringer, W., *Witchcraft Persecutions in Bavaria* (Cambridge: Cambridge University Press, 1997).

—, 'Climatic Change and Witch-Hunting: The Impact of the Little Ice Age on Mentalities', *Climatic Change*, 43 (1999), pp. 335–51.

—, 'Die Krise von 1570: Ein Beitrag zur Krisengeschichte der Neuzeit', in M. Jakubowski-Tiessen and H. Lehmann (eds), *Um Himmels Willen: Religion in Katastrophenzeiten* (Göttingen: Vandenhoeck & Ruprecht, 2003), pp. 51–156.

Berg, H., *Military Occupation under the Eyes of the Lord: Studies in Erfurt during the Thirty Years War* (Göttingen: Vandenhoeck & Ruprecht, 2010).

Bischoff, M., *Himmelszeichen: Eine bildreiche Kunde von Aberglauben und Ängsten* (Nördlingen: Franz Greno, 1986).

Boettcher, S., 'The Rhetoric of "Seelsorge" for Miners in the Sermons of Cyriakus Spangenberg', in G. Litz, H. Munzert and R. Liebenberg (eds), *Frömmigkeit-Theologie-Frömmigkeitstheologie: Contributions to European CH Festschrift für Berndt Hamm* (Leiden: Brill, 2005), pp. 453–66.

Bossert, G., 'Irenaeus, Christoph', in P. Schaff (ed.), *The New Schaff-Herzog Encyclopedia of Religious Knowledge*, 13 vols (Grand Rapids, MI: Baker Books, 1953), vol. 6, p. 31.

Brazdil, R., et al., 'Flood Events of Selected European Rivers in the Sixteenth Century', in C. Pfister, R. Brazdil and R. Glaser (eds), *Climatic Variability in Sixteenth-Century Europe and its Social Dimension* (Dordrecht: Kluwer, 1999), pp. 252–61.

Caroti, S., 'Melanchthon's Astrology', in P. Zambelli (ed.), *'Astrologi hallucinati': Stars and the End of the World in Luther's Time* (Berlin: Walter de Gruyter, 1986), pp. 109–21.

Christensen, C. C., *Princes and Propaganda: Electoral Saxon Art of the Reformation* (Kirksville, MO: Sixteenth Century Journal Publishers, 1992).

Christman, R., 'Heretics in Luther's Homeland: The Controversy over Original Sin in Late Sixteenth-Century Mansfeld' (PhD dissertation, University of Arizona, 2004).

—, *Doctrinal Controversy and Lay Religiosity in Late Reformation Germany: The Case of Mansfeld* (Leiden: Brill, 2012).

Conner, J. A., *Kepler's Witch: An Astronomer's Discovery of Cosmic Order amid Religious War, Political Intrigue, and the Heresy Trial of his Mother* (San Francisco, CA: HarperSanFrancisco, 2004).

Crawford, J., *Marvelous Protestantism: Monstrous Births in Post-Reformation England* (Baltimore, MD: Johns Hopkins University Press, 2005).

Deneke, B., 'Kaspar Goltworm: Ein lutherischer Kompilator zwischen Überlieferung und Glaube', in W. Brückner (ed.), *Volkserzählung und Reformation: Ein Handbuch zur Tradierung und Funktion von Erzählstoffen und Erzählliteratur im Protestantismus* (Berlin: Erich Schmidt, 1974), pp. 125–77.

Dickson, D. R., 'Johann Valentin Andreae's Utopian Brotherhoods', *Renaissance Quarterly*, 49:4 (1996), pp. 760–98.

Dingel, I., *Concordia Controversa: Die öffentlichen Diskussionen um das lutherische Konkordienwerck am Ende des 16. Jahrhunderts* (Gütersloh: Gütersloher Verlagshaus, 1996).

—, 'The Culture of Conflict in the Controversies Leading to the Formula of Concord (1548–1580)', in R. Kolb (ed.), *Lutheran Ecclesiastical Culture* (Leiden: Brill, 2008), pp. 15–64.

Dixon, C. S., 'Popular Astrology and Lutheran Propaganda in Reformation Germany', *History*, 84 (1999), pp. 403–18.

Duke, A., 'A Legend in the Making: News of the "Spanish Inquisition" of the Low Countries in German Evangelical Pamphlets, 1546–1550', *Nederlandsch archief voor kerkgeschiedenis*, 22 (1997) pp. 125–44.

Fleischer, M. P., 'The Reception in Silesia' in L. W. Spitz and W. Lohff (eds), *Discord, Dialogue, and Concord: Studies in the Lutheran Reformation's Formula of Concord* (Philadelphia, PA: Fortress Press, 1977), pp. 119–35.

Francisco, A. S., *Martin Luther and Islam: A Study in Sixteenth-Century Polemics and Apologetics* (Leiden: Brill, 2007).

Frank, G., 'Irenaeus, Christoph', in *Allgemeine Deutsche Biographie*, 56 vols (Leipzig: Duncker & Humblot, 1881), vol. 14, p. 582.

Friedman, J., *The Battle of the Frogs and Fairford's Flies: Miracles and the Pulp Press during the English Revolution* (New York: St Martin's Press, 1993).

Frymire, J. M., *The Primacy of the Postils: Catholics, Protestants, and the Dissemination of Ideas in Early Modern Germany* (Leiden: Brill, 2010).

Goodale, J., 'Intimidation, Intolerance, and Injury: Religious Conflict in Saxony, 1587–1592', in K. von Greyerz and K. Siebenhüner (eds), *Religion und Gewalt: Konflikte, Rituale, Deutungen (1500–1800)* (Göttingen: Vandenhoeck & Ruprecht, 2006), pp. 195–219.

Haemig, M. J., 'The Living Voice of the Catechism: German Catechetical Preaching 1530–1580' (PhD dissertation, Harvard Divinity School, 1996).

—, 'Laypeople as Overseers of the Faith: A Reformation Proposal', *Trinity Seminary Review*, 27:1 (2006), pp. 21–7.

Hammer, W., 'Melanchthon, Inspirer of the Study of Astronomy', *Popular Astronomy*, 59 (1951), pp. 308–19.

Hartfelder, K., 'Der Aberglaube Philipp Menlanchthons', *Historisches Taschenbuch*, 6 (1889), pp. 233–69.

Hellmann, G., *Die Meteorologie in den deutschen Flugschriften und Flugblättern des XVI Jahrhunderts* (Berlin: Preussischen Akademie der Wissenschaften, 1921).

Hoffman-Randall, C., *Monster, Wunder und Kometen: Sensationsberichte auf Flugblätterm des 16. bis 18. Jahrhunderts* (Erlangen: Univeritäts Bibliothek, 1999).

Homeyer, S., 'Die kompensatorische Funktion für den Rezipenten bei einigen illustrierten Flugblättern des 16. Jahrhunderts mit Endzeitvorstellungen', in W. Harms and M. Schilling (eds), *Das Illustrierte Flugblatt in der Kultur der Frühen Neuzeit* (Frankfurt am Main: Peter Lang, 1998), pp. 137–49.

—, '"...das ende mus verhanden sein!...": Studien zur eschatologischen Bildlichkeit auf illustrierten Flugblättern der Frühen Neuzeit' (PhD dissertation, Otto-von-Guericke University, 2002).

Ilic, L., 'The Understanding of Sin in the Theology of Matthias Flacius', *Perichoresis*, 4:2 (2006), pp. 211–44.

Janeck, A., *Zeichen am Himmel: Flugblätter des 16. Jahrhunderts* (Nuremberg: Germanisches National Museum, 1982).

Jansen, S. L., *Political Protest and Prophecy under Henry VIII* (Woodbridge: Boydell Press, 1991).

Jervis, J. L., *Cometary Theory in Fifteenth-Century Europe* (Dordrecht: D. Reidel, 1985).

Junghans, H., 'Irenäus, Christoph', in *Neue Deutsche Biographischie*, 25 vols (Berlin: Duncker & Humblot, 1974), vol. 10, p. 178.

Jungkuntz, T. R., *Formulators of the Formula of Concord: Four Architects of Lutheran Unity* (St Louis, MO: Concordia, 1977).

Karant-Nunn, S. C., 'Neoclericalism and Anticlericalism in Saxony, 1555–1675', *Journal of Interdisciplinary History*, 29:4 (1994), pp. 615–37.

Kaufmann, T., *Ende der Reformation: Magdeburgs 'Herrgottes Kanzlei (1548–1551/2)* (Tübingen: Mohr Siebeck, 2003).

—, '"Our Lord God's Chancery" in Magdeburg and its Fight against the Interim', *Church History*, 73:3 (2004), pp. 566–82.

—, 'The Lutheran Sermon during the War and at the Time of the Peace Agreement', in K. Bussmann and H. Schilling (eds), *1648: War and Peace in Europe*, 3 vols (Münster: Westfälisches Landesmuseum, 2008), vol. 1, pp. 245–50.

Kern, E. M., 'The "Universal" and the "Local" in Episcopal Visitations', in M. Reinhart (ed.), *Infinite Boundaries: Order, Disorder, and Reorder in Early Modern German Culture* (Kirksville, MO: Sixteenth Century Journal Publishers, 1998), pp. 35–54.

Klug, E. F., *Getting into the Formula of Concord: A History and Digest of the Formula* (St Louis, MO: Concordia, 1977).

Kolb, R., 'Nikolaus von Amsdorf, Knight of God and Exile of Christ: Piety and Polemic in the Wake of Luther' (PhD dissertation, University of Wisconsin, 1973).

—, 'Parties, Princes, Pastors, and Peace: The Formulation of Concord, 1577', *Academy: Lutherans in Profession*, 34 (1977), pp. 2–13.

—, 'Jakob Andrae's Concern for the Laity', *Concordia Journal*, 4 (1978), pp. 58–67.

—, 'The Flacian Rejection of the Concordia, Prophetic Style and Action on the German Late Reformation', *Archiv für Reformationgeschichte*, 73 (1982), pp. 196–216.

—, *Confessing the Faith: Reformers Define the Church, 1530–1580* (St Louis, MO: Concordia, 1991).

—, 'Philipp Melanchthon: Reformer and Theologian', *Concordia Journal*, 23:4 (1997), pp. 309–16.

—, *Martin Luther as Prophet, Teacher, and Hero: Images of the Reformer, 1520–1620* (Grand Rapids, MI: Baker Books, 1999).

—, 'The Legal Case for Martyrdom: Basilius Monner on Johann Friedrich the Elder and the Smalcald War', in I. Dingel, V. Leppin and C. Strohm (eds), *Reformation und Recht* (Gütersloh: Chr. Kaiser, 2002), pp. 145–60.

—, and J. Nestingen (eds), *Sources and Contexts of the Book of Concord* (Minneapolis, MN: Fortress Press, 2001).

Koslofsky, C., *The Reformation of the Dead: Death and Ritual in Early Modern Germany, 1450–1700* (New York: St Martin's Press, 2000).

Kreitzer, B., 'The Lutheran Sermon', in L. Taylor (ed.), *Preachers and People in the Reformations and Early Modern Period* (Leiden: Brill, 2001), pp. 35–63.

Krusenstjern, B. von, 'Prodigienglaube und Dreißigjähriger Krieg', in H. Lehmann and A.-C. Trepp (eds), *Zeichen der Krise: Religiosität im Europa des 17. Jahrhunderts* (Göttingen: Vandenhoeck & Ruprecht, 1999), pp. 53–78.

Kurihara, K., '"Whether We Live or Die, We Are the Lord's": Lutheran *Wetter* Discourses and Spiritual Interpretations of Natural Disasters', in K. A. Edwards (ed.), *The Science of the Supernatural in Early Modern Europe* (New York: Palgrave Macmillan, forthcoming).

Kusukawa, S., 'Aspectio divinorum operum: Melanchthon and Astrology for Lutheran Medics', in O. P. Grell and A. Cunningham (eds), *Medicine and the Reformation* (New York: Routledge, 1993), pp. 33–56.

Leppin, V., *Antichrist und Jüngster Tag: das Profil apokalyptischer Flugschriftenpublizistik im deutschen Luthertum* (Gütersloh: Gütersloher Verlagshaus, 1999).

Lohrmann, K., 'Irenäus, Christoph', in *Biographisch-Bibliographisches Kirchenlexikon*, 14 vols (Hamm: Traugott Bautz, 1990), vol. 2, pp. 1327–8.

Manschreck, C. L., 'The Role of Melanchthon in the Adiaphora Controversy', *Archiv für Reformationsgeschichte*, 48 (1957), pp. 165–82.

Martin, F., *Hellbrunn Castle near Salzburg* (Salzburg: Salzburger Verlag für Wirtschaft und Kultur, 1957).

Mauelshagen, F., 'Illustrierte Kometenflugblätter in wahrnehmungsgeschichtlicher Perspektive', in W. Harms and M. Schilling (eds), *Das Illustrierte Flugblatt in der Kultur der Frühen Neuzeit* (Frankfurt am Main: Peter Lang, 1998), pp. 101–36.

Maunder, M., *Lights in the Sky: Identifying and Understanding Astronomical and Meteorological Phenomena* (London: Springer, 2007).

Messerli, A., 'Angst und Wunderzeichen in Einblattdrucken', *Librarium*, 31 (1988), pp. 182–97.

Methuen, C., 'The Role of the Heavens in the Thought of Philip Melanchthon', *Journal of the History of Ideas*, 57:3 (1996), pp. 385–403.

—, '"This Comet or New Star": Theology and the Interpretation of the Nova of 1572', *Perspectives on Science*, 5 (1997), pp. 499–514.

—, *Kepler's Tübingen: Stimulus to a Theological Mathematics* (Aldershot: Ashgate, 1998).

Miller, G. J., 'Holy War and Holy Terror: Views of Islam in German Pamphlet Literature, 1520–1545' (PhD dissertation, Boston University, 1994).

Montgomery, J. W., 'Cross, Constellation, and Crucible: Lutheran Astrology and Alchemy in the Age of the Reformation', *Ambix*, 11 (1963), pp. 65–86.

Mortimer, G., *Eyewitness Accounts of the Thirty Years War 1618–48* (New York: Palgrave, 2002).

Naylor, J., *Out of the Blue: A 24-Hour Skywatcher's Guide* (Cambridge: Cambridge University Press, 2002).

Nischan, B., 'The Palatine and Brandenburg's "Second Reformation"', in D. Visser (ed.), *Controversy and Conciliation: The Reformation and the Palatine 1559–1583* (Allison Park, PA: Pickwick Publications, 1986), pp. 303–18.

—, *Lutherans and Calvinists in the Age of Confessionalism* (Aldershot: Ashgate, 1999).

Oettinger, R. W., *Music as Propaganda in the German Reformation* (Aldershot: Ashgate, 2001).

Olson, O. K., 'Theology of Revolution: Magdeburg, 1550–1551', *Sixteenth Century Journal*, 3:1 (1972), pp. 56–79.

—, *Matthias Flacius and the Survival of Luther's Reform* (Wiesbaden: Harrassowitz, 2002).

Parker, G., *The Thirty Years' War* (New York: Military Heritage Press, 1987).

—, 'Success and Failure during the First Century of the Reformation', *Past and Present*, 136:1 (1992), pp. 43–82.

Pfister, C., and R. Brazdil, 'Climatic Variability in Sixteenth-Century Europe and its Social Dimension: A Synthesis', in C. Pfister, R. Brazdil and R. Glaser (eds), *Climatic Variability in Sixteenth-Century Europe and its Social Dimension* (Dordrecht: Kluwer, 1999), pp. 5–53.

Piepkorn, A. C., 'Martin Chemnitz's View on Trent: The Genesis and the Genius of the *Examen Concolii Tridentini*', *Concordia Theological Monthly*, 37 (1966), pp. 5–37.

Plakans, A., *Historical Dictionary of Latvia* (Lanham, MD: Scarecrow Press, 1997).

Rauscher, J., 'Der Halleysche Komet im Jahre 1531 und die Reformatoren', *Zeitschrift für Kirchengeschichte*, 32 (1911), pp. 259–76.

Rein, N., *The Chancery of God: Protestant Print, Polemic and Propaganda against the Empire, Magdeburg 1546–1551* (Aldershot: Ashgate, 2008).

Rittgers, R. K., *The Reformation of Suffering: Pastoral Theology and Lay Piety in Late Medieval and Early Modern Germany* (Oxford: Oxford University Press, 2012).

Roos, K. L., *The Devil in 16th Century German Literature: The Teufelsbucher* (Bern: Herbert Lang, 1972).

Savage, C., *Aurora: The Mysterious Northern Lights* (San Francisco, CA: Sierra Club Books, 1994).

Schenda, R., 'Die deutschen Prodigiensammlungen des 16. und 17. Jahrhunderts', *Archiv für Geschichte des Buchwesens*, 4 (1963), pp. 637–710.

—, 'Irenaeus, Christoph', in *Enzyklopädie des Märchens*, 14 vols (Berlin: Walter de Gruyter, 1991), vol. 7, pp. 270–4.

—, 'Wunder-Zeichen: Die alten Prodigen in neuen Gewändern: Eine Studie zur Geschichte eines Denkmusters', *Fabula*, 38 (1997), pp. 14–32.

Schilling, H., 'Job Fincel und die Zeichen der Endzeit', in W. Brückner (ed.), *Volkserzählung und Reformation: Ein Handbuch zur Tradierung und Funktion von Erzählstoffen und Erzählliteratur im Protestantismus* (Berlin: Erich Schmidt, 1974), pp. 326–93.

Schilling, M., 'Flugblatt und Krise in der Frühen Neuzeit', in W. Harms and A. Messerli (eds), *Wahrnehmungsgeschichte und Wissensdiskurs im illustrierten Flugblatt der Frühen Neuzeit (1450–1700)* (Basel: Schwabe, 1999), pp. 33–56.

Schmidt-Biggemann, W., 'The Apocalypse and Millenarianism in the Thirty Years' War', in K. Bussmann and H. Schilling (eds), *1648: War and Peace in Europe*, 3 vols (Münster: Westfälisches Landesmuseum, 1998), vol. 1, pp. 259–63.

Schofield, J., *Philip Melanchthon and the English Reformation* (Aldershot: Ashgate, 2006).

Schwegler, M., *'Erschröckliches Wunderzeichen' oder 'natürliches Phänomenon'?: Frühneuzeitliche Wunderzeichenberichte aus der Sicht der Wissenschaft* (Munich: Percy Berktold, 2002).

Scribner, R. 'Anticlericalism and the German Reformation', in *Popular Culture and Popular Movements in Reformation Germany* (London: Hambledon Press, 1987), pp. 243–56.

Soergel, P. M., 'Luther on Angels', in P. Marshall and A. Walsham (eds), *Angels in the Early Modern World* (Cambridge: Cambridge University Press, 2006), pp. 64–82.

—, *Miracles and the Protestant Imagination: The Evangelical Wonder Book in Reformation Germany* (Oxford: Oxford University Press, 2012).

Spinks, J., *Monstrous Births and Visual Culture in Sixteenth-Century Germany* (London: Pickering & Chatto, 2009).

—, 'Print and Polemic in Sixteenth-Century France: The *Histories prodigieuses*, Confessional Identity, and the Wars of Religion', *Renaissance Studies* (2011), pp. 1–25.

Spitz, L. W., 'The Formula of Concord Then and Now', in L. W. Spitz and W. Lohff (eds), *Discord, Dialogue, and Concord: Studies in the Lutheran Reformation's Formula of Concord* (Philadelphia, PA: Fortress Press, 1977), pp. 1–12.

Starenko, P. E., 'In Luther's Wake: Duke John Frederick II of Saxony, Angelic Prophecy, and the Gotha Rebellion of 1567' (PhD dissertation, University of California Berkeley, 2002).

Strauss, G., 'Success and Failure in the German Reformation', *Past and Present*, 67:1 (1975), pp. 30–63.

—, *Luther's House of Learning: Indoctrination of the Young in the German Reformation* (Baltimore, MD: Johns Hopkins University Press, 1978).

—, 'Local Anticlericalism in Reformation Germany', in P. A. Dykema and H. A. Oberman (eds), *Anticlericalism in Late Medieval and Early Modern Europe* (Leiden: Brill, 1993), pp. 625–37.

Taylor, R., *The Political Prophecy in England* (New York: Columbia University Press, 1911).

Theibault, J., 'Jeremiah in the Village: Prophecy, Preaching, Pamphlets, and Penance in the Thirty Years' War', *Central European History*, 27:4 (1994), pp. 441–60.

Thornedike, L., *A History of Magic and Experimental Science*, 5 vols (New York: Columbia University Press, 1941).

Tlusty, B. A., *The Martial Ethic in Early Modern Germany: Civic Duty and the Right of Arms* (New York: Palgrave Macmillan, 2011).

Wallmann, J., 'Johann Arndt (1555–1621)', in C. Lindberg (ed.), *The Pietist Theologians: An Introduction to Theology in the Seventeenth and Eighteenth Centuries* (Malden: Blackwell, 2005), pp. 21–37.

Walsham, A., *Providence in Early Modern England* (Oxford: Oxford University Press, 1999).

—, 'Sermons in the Sky: Apparitions in Early Modern Europe', *History Today*, 51:4 (2001), pp. 56–63.

Weber, B., *Wunderzeichen und Winkeldrucker 1543–1586* (Zurich: Urs Graf, 1972).

Wengert, T. J., '"Not by Nature *Philoneikos*": Philip Melanchthon's Initial Reactions to the Augsburg Interim', in *Philip Melanchthon, Speaker of the Reformation* (Farnham: Ashgate, 2010), Essay XIII, pp. 33–49.

Whitford, D. M., *Tyranny and Resistance: The Magdeburg Confession and the Lutheran* Tradition (St Louis, MO: Concordia, 2001).

Wriedt, M., 'Between Angst and Confidence: Melanchthon as a Man of the Sixteenth Century', *Concordia Journal*, 23:4 (1997), pp. 277–94.

Yates, F. A., *The Rosicrucian Enlightenment* (London: Routledge, 1971).

Zika, C., *The Appearance of Witchcraft: Print and Visual Culture in Sixteenth-Century Europe* (London and New York: Routledge, 2007).

INDEX

For Product Safety Concerns and Information please contact our EU representative GPSR@taylorandfrancis.com
Taylor & Francis Verlag GmbH, Kaufingerstraße 24, 80331 München, Germany

www.ingramcontent.com/pod-product-compliance
Ingram Content Group UK Ltd.
Pitfield, Milton Keynes, MK11 3LW, UK
UKHW021303190926
475701UK00009B/627
* 9 7 8 1 1 3 8 6 6 3 0 6 0 *